Power Electronics: Emerging Technologies and Applications

Power Electronics: Emerging Technologies and Applications

Edited by **Maurice Willis**

CWILLFORD PRESS

New York

Published by Willford Press,
118-35 Queens Blvd., Suite 400,
Forest Hills, NY 11375, USA
www.willfordpress.com

Power Electronics: Emerging Technologies and Applications
Edited by Maurice Willis

International Standard Book Number: 978-1-68285-085-5 (Hardback)

Printed in the United States of America.

Contents

Preface

There has been rapid advancement in the field of power electronics. The emphasis is on developing innovative power systems that are energy efficient and can be utilized for longer durations. This book includes topics that provide a comprehensive overview of emerging applications like nanoelectronics, bioelectronics, low power processors, etc. It includes case studies that highlight the advanced applications of power electronics in different fields like bio-medical science, sensor technology, multimedia, etc. The extensive content of this book will help the students and academicians in gaining a thorough understanding of the field.

This book is a comprehensive compilation of works of different researchers from varied parts of the world. It includes valuable experiences of the researchers with the sole objective of providing the readers (learners) with a proper knowledge of the concerned field. This book will be beneficial in evoking inspiration and enhancing the knowledge of the interested readers.

In the end, I would like to extend my heartiest thanks to the authors who worked with great determination on their chapters. I also appreciate the publisher's support in the course of the book. I would also like to deeply acknowledge my family who stood by me as a source of inspiration during the project.

Editor

Analog Encoding Voltage—A Key to Ultra-Wide Dynamic Range and Low Power CMOS Image Sensor

Arthur Spivak [1,*], Alexander Belenky [1], Alexander Fish [2] and Orly Yadid-Pecht [3]

[1] The VLSI Systems Center, LPCAS, Ben-Gurion University, P.O.B. 653, Be'er-Sheva 84105, Israel;
E-Mail: belenky@ee.bgu.ac.il

[2] Department of Electrical and Computer Engineering, Bar Ilan University, Ramat Gan, 52100, Israel;
E-Mail: alexander.fish@biu.ac.il

[3] Department of Electrical Engineering, University of Calgary, Alberta T2N 1N4, Canada;
E-Mail: orly.yadid.pecht@ucalgary.ca

* Author to whom correspondence should be addressed; E-Mail: spivakar@post.bgu.ac.il

Abstract: Usually Wide Dynamic Range (WDR) sensors that autonomously adjust their integration time to fit intra-scene illumination levels use a separate digital memory unit. This memory contains the data needed for the dynamic range. Motivated by the demands for low power and chip area reduction, we propose a different implementation of the aforementioned WDR algorithm by replacing the external digital memory with an analog in-pixel memory. This memory holds the effective integration time represented by analog encoding voltage (*AEV*). In addition, we present a "ranging" scheme of configuring the pixel integration time in which the effective integration time is configured at the first half of the frame. This enables a substantial simplification of the pixel control during the rest of the frame and thus allows for a significantly more remarkable DR extension. Furthermore, we present the implementation of "ranging" and *AEV* concepts on two different designs, which are targeted to reach five and eight decades of DR, respectively. We describe in detail the operation of both systems and provide the post-layout simulation results for the second solution. The simulations show that the second design reaches DR up to 170 dBs. We also provide a comparative analysis in terms of the number of operations per pixel required by our solution and by other widespread WDR algorithms. Based on the calculated results, we conclude that the proposed two designs, using "ranging" and *AEV* concepts, are attractive, since they obtain a wide dynamic range at high operation speed and low power consumption.

Keywords: CMOS; image sensor; low power; rolling shutter; snapshot; SNR; wide dynamic range

1. Introduction

Extending the dynamic range (DR) of a CMOS image sensor's (CIS) remains one of the challenges to be faced in designing an effective versatile sensor. Improvement of image capture capability can be done either by reducing the noise floor (NF) of the sensor [1–3] or by extending its saturation toward higher light intensities. Most of the cases that are reported in the literature give solutions which focus on extending the DR toward high light intensities. There are numerous solutions that have been proposed throughout the years [4–27]. However, the majority of the proposed WDR solutions have used an external memory, which significantly increased their area and power consumption and, as a result, their cost.

Examples of using a digital memory can be found in a large variety of solutions [8] such as (a) multimode sensors that have a linear and a logarithmic response at dark and bright illumination levels, respectively [9]; (b) clipping sensors, in which a well capacity adjustment method is applied [10]; (c) frequency-based sensors, in which the sensor output is converted into a pulse frequency [12]; (d) time-to-first spike (TFS) sensors, in which the image is processed according to the time the pixel was detected as saturated [9]; (e) sensors with global control over the integration time, in which the pixel integrates for a set of exposures regardless of its input light intensity [15]; and (f) sensors with autonomous control over the integration time (multiple resets), in which each pixel has control over its own exposure period [16–18]. In the aforementioned designs the memories were used to store the possible pixel outputs or the DR extension data. The reason for using a memory is that the data processing becomes more straightforward, *i.e.*, it is easy to split the processing into several independent stages and to store the intermediate results. However, in such a case, the overall processing time and power increase.

Nevertheless, we can find memory free WDR solutions as well. For example, in [19] a TFS algorithm implementation is described, where, instead of using a digital code to represent the saturation time, the authors paired each time slot with a certain voltage value that was subsequently converted into corresponding digital pixel value. This solution unfortunately suffers from the reduced sensitivity throughout the whole DR [8].

It is especially interesting to look into the free memory solutions in [20,21], which are based upon the global control over the integration time, since this category is the most widely spread in the industry. The memory can be omitted if there are two exposures as will be shown later on. However, using two exposures causes a substantial signal to noise ratio (SNR) drop in the mid-tones, which can necessitate subsequent image processing.

With autonomous control over the integration time category, the memory emerges as the main cause for the increase in both the complexity and the cost of the solution. Attempts to dump the memory in this category have been undertaken as well. In [22] a successful memory free solution was presented. That solution was based upon a very simple pixel structure, in which the pixel integration time value was encoded in voltage and stored on the intra-pixel capacitance. But, since the same capacitance was also used for readout of the intermediate pixel values, the stored integration times were continuously overwritten and refreshed over and over again slowing the sensor operation.

In this work, we present two novel concepts in "ranging" and analog encoding voltage (*AEV*) for design of low power, high speed, memory free WDR sensors with autonomous control over integration time. According to the "ranging" concept, the pixel signal is "coarsely" quantized at the beginning of the frame, and the final "fine" quantization takes place at the end of the frame. Generated data in the "coarse" quantization is stored as analog encoding voltage (*AEV*) inside each pixel on a separate capacitance, and this controls the pixel integration throughout the frame. Since we use different capacitances for a photo-induced signal and for an (*AEV*), the data processing becomes very simple and fluent. At the end of the frame, both of the two analog values, the photo-generated charge and the *AEV*, are merged through an analog to digital A/D conversion to a final digitized pixel value.

In addition, we present two designs which implement the "ranging" and *AEV* concepts. The designs are thoroughly described from the schematic to the layout. Moreover, we present the post-layout simulation results of the second design, showing successful DR extension up to 170 dBs. To illustrate the effectiveness of the proposed solutions, we perform a qualitative comparison with other published solutions and show that eventually our solutions are optimized for low power WDR imaging.

The presented work is organized as follows: Section 2 introduces our solutions, Section 3 presents the comparison of a number of operations in several types of WDR solutions, and Section 4 concludes the study. Following the last section, there are two Appendixes, namely A and B. They include the detailed considerations of pixel swing segmentation and signal to noise (SNR) considerations, respectively. In this way, the overall concept becomes clearer and no important design consideration is omitted.

2. WDR Solution Based upon Ranging and AEV

2.1. Problem Definition and Proposed Solution

The solution we have used for the DR extension is called the autonomous control over the integration time (multiple resets). None of the pixels in this solution are classic 3T's, because this enables individual pixel reset via an additional transistor [8]. During the frame the outputs of a selected row are read nondestructively through the regular output chain. Then they are compared with an appropriate threshold at certain time points by comparators, which are found at a separate unit outside the pixel array. If a pixel value exceeds the threshold, a reset is given at that time point to that pixel. The binary information concerning the reset being applied or not is saved in a digital storage, in order to enable proper scaling of the value read. This enables the pixel value to be described as a floating-point representation, wherein the exponent will describe the scaling factor for the actual integration time, while the mantissa will be the regular A/D output. In this way the actual pixel value would be:

$$V_{pixel} = M_{pixel} X^{EXP}$$

(1)

where V_{pixel} is the actual pixel value, M_{pixel} is the analog or digitized output value that has been read out at the end of the frame, X is a chosen constant, 2, for instance. EXP is the exponent value, which describes the scaling factor, *i.e.*, the specific part of the integration time the pixel actually integrated without being saturated.

The bottleneck of the basic solution is that bits representing the EXP are generated during the integration sequentially; hence, after the generation of the subsequent bit, a digital memory should be refreshed. As the number of pixel rows increase, the time that can be allocated to EXP bits' generation becomes more limited. The restrictions on the DR extension vary according to the mode of sensor operation: rolling or global shutter. This is more obvious in the rolling shutter, wherein the integration in the successive rows is skewed, so there is more time for DR processing. On the other hand, in the global shutter sensor, wherein the integration starts simultaneously in all rows, the processing times are substantially reduced. Therefore, the solution that we propose is targeted mainly for the global shutter sensor.

2.2. The Proposed Solution

Our solution is based upon "ranging" and AEV concepts. The purpose of this solution is to reduce DR processing times and to omit the memory unit. In addition, we have presented the solution architecture and the flow.

We have divided the WDR algorithm into three phases (Figure 1). During Phase 1, the EXP values are produced and memorized inside each pixel in a format of analog encoding voltage AEV. This assignment, which precedes the saturation checks, is called "ranging", since it "coarsely" quantizes [23] the pixel signal. In fact, the AEV value produced in this quantization, contains the most significant bits of pixel signal $Pix<M + L - 1:L>$ (Figure 2); whereas the *Mantissa* obtained at the end of the next phase contains the least significant bits $Pix<L - 1:0>$. In this way, each pixel is given its valid integration time for the current frame before the saturation checks start.

Figure 1. Three Phases of the proposed WDR algorithm.

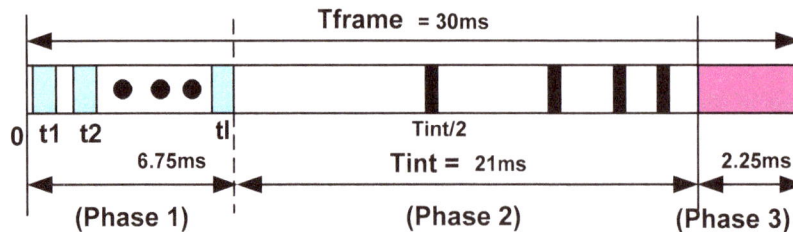

During Phase 2, the pixel integrates in accordance with its AEV. Since the integration time is already known, there is no need to retrieve the last reset information or to refresh information for the current check as was performed in our previous solutions; we merely compare the AEV to the global reference. In case the *AEV* is below the reference, the pixel will be reset; if not, the pixel will remain untouched till the next frame.

Figure 2. Digital Pixel Value: Mantissa and the EXP bits.

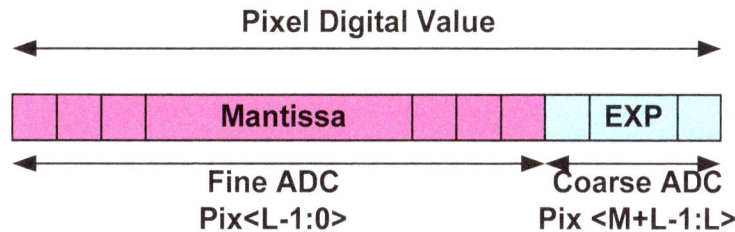

During Phase 3, both the *AEV* and *Mantissa* are digitized in a single slope analog to a digital conversion (ADC) (Figure 3). For this purpose, the *AEV* and *Mantissa* are read separately and converted with different resolutions. The *AEV* is digitized with a "coarse" resolution; whereas the *Mantissa* is converted by a ramp with a "fine" resolution.

Figure 3. The Coarse and Fine Quantization of the Pixel Signal.

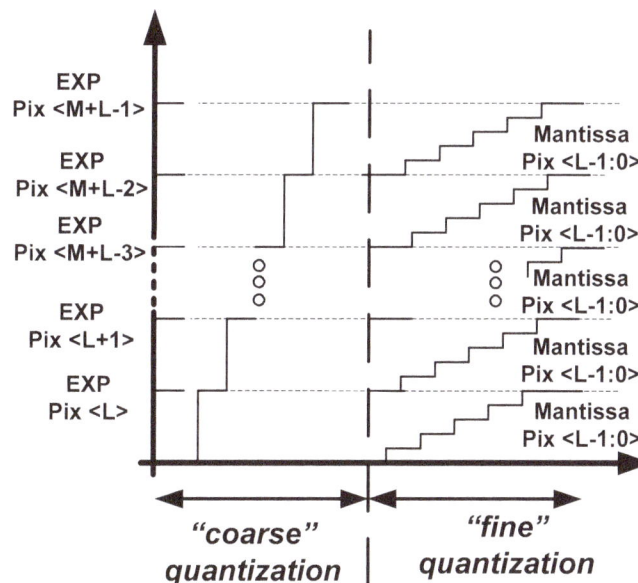

In general, the three phases can overlap each other. However, in this study, we present the first version of the algorithm; therefore, we assume that each phase is completely separated from the others. In future works we will demonstrate how these phases can be overlaid on each other. We assessed the duration of each phase in accordance with further presented simulation results. In Figure 1, we present the division of a single frame (30 ms) into three phases as if the algorithm is applied on a 1000 × 1000 pixel array. In spite of all the phases being separate, it is important to note that the maximal integration time (Phase 2) still occupies the major part of the frame, whereas the overall integration times in the rest of the phases 1 and 3 are substantially shorter.

2.3. Phase 1: Assigning the Analog Encoding Voltage (AEV)

How are the values of *AEV* selected and paired with the corresponding *EXP* bits? The *AEV* values can be dispersed within the voltage domain in several ways: logarithmically or linearly (Figure 4). In this study we chose, for the sake of convenience, a linear quantization, according to which the all the *AEV* values are spaced at the same distance of $2\Delta V_{min}$. We also decided that the *AEV* corresponding to the highest light intensity would be represented by the lowest available voltage, since this simplifies all the subsequent conversions. The parameter ΔV_{min} is derived from the properties of the comparator that are to be employed for assigning the *AEV* values. The requirements for ΔV_{min} are intuitive and straightforward:

$$\Delta V_{min} \geq \frac{V_{out\_comparator}}{G}, \quad \Delta V_{min} \geq V_{offset} \tag{2}$$

in which $V_{out\_comparator}$, G, V_{offset} are the comparator's output swing, gain, and offset, respectively.

Figure 4. Analog encoding voltage (*AEV*) concept: each dynamic range (DR) bit has its own Analog Encoding Voltage.

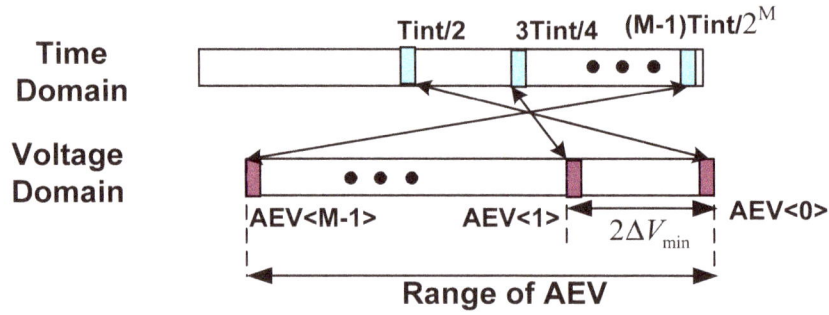

After the pixel array has been reset globally, Phase 1 begins. The pixels are exposed to light and accumulate the photo-generated charge. At certain predetermined time points, each pixel row is sampled nondestructively and spanned by a piecewise increasing ramp. In case the pixel signal intersects with the ramp, the comparator flips and the corresponding *AEV* is written into the in-pixel dynamic memory. Detailed considerations of the *AEV* assignment are provided in Appendix A.

2.4. Phase 2 Performing the Conditional Resets

Phase 2 performs the conditional resets to the pixel array. Since the base of the DR extension is 2, the integration times are ordered in exactly the same way as in our previous solutions [24], *i.e.*, in geometric descending sequence:

$$\frac{T_{int}}{2}, \frac{T_{int}}{2^2} \cdots \frac{T_{int}}{2^{EXP}} \tag{3}$$

Nonetheless, in the current solution, the pixel reset cycle can be reduced by a factor of 2 at least, since the number of operations before the conditional reset is halved. All that has to be done is to compare the individual *AEV* with the reference, which corresponds to the current reset cycle. This phase can be performed either in a traditional fashion, *i.e.*, row by row scan, or instantly within the whole array. We will discuss in further detail two different designs implementing both of the options, namely **TYPE I** and **TYPE II**.

2.5. Phase 3: Coarse, Fine ADC & Readout

Phase 3 concludes the frame. At this stage, the analog pixel value (*Mantissa*) and the *AEV* are converted by the same comparators, which had been used in the two preceding phases. In fact, we performed a single slope ADC to obtain the digitized values of *Mantissa* and *AEV* (Figure 2). The conversion is performed row by row: first, the *AEV* is spanned with the "coarse" resolution, and then the pixel swing is spanned by the ramp with "fine" resolution. Of course, after the span, both the *AEV* and pixel capacitances are reset to their initial predetermined values.

2.6. TYPE I Pixel

For most applications the required dynamic range (DR) does not exceed 120 dBs, so the pixel structure can be straightforward. In our case, all we need is to accommodate the *AEV* and the pixel *Mantissa* on two separate capacitors and to add a readout chain and conditional access circuitry (Figure 5) to each of them. Since the pixel structure here is not complex and the DR extension is typical, this design is named **TYPE I**.

Figure 5. Schematic of **TYPE I** Pixel.

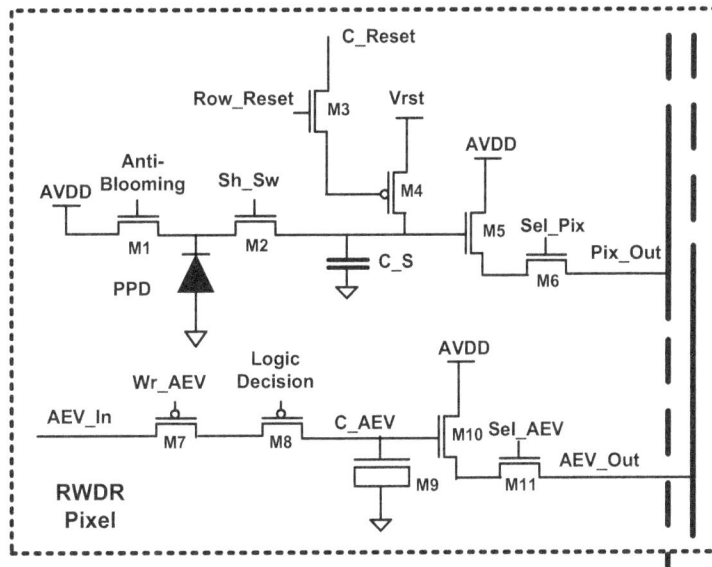

The functionality of the **TYPE I** pixel is implemented with 11 transistors only. They can be divided into two groups: the first processes the data received from the pinned photodiode (*PPD*) M_1–M_6; and the second is responsible for the *AEV* handling M_7–M_{11}.

The charge transfer from the *PPD* is bidirectional, similar to [25]. In this way, we can deliberately dump the generated charge to drain the photo-diode through M_1 or we can transfer the charge for further processing to C_s through M_2. We use a simple conditional reset scheme, implemented by M_3 and M_4 transistors, to implement the multiple resets algorithm. By activating the *Row_Reset* signal, the required pixel row is chosen and then, by means of the *C_Reset* signal, driven by the column-wise comparators, is conditionally reset to V_{rst}. Transistors M_5, M_6 form a traditional source follower to handle the C_S readout [26].

AEV processing inside the pixel is very simple as well. The *AEV* value is fed to the pixel from the column-wise bus *AEV_In*. The write operation of the input *AEV* into the pixel dynamic memory is conditional and is implemented by using a trivial stacked scheme M_7 and M_8. By activating the *Wr_AEV* signal, a required row is accessed; while the *Logic Decision* signal, driven by column-wise comparators, enables the MOS capacitor C_{AEV} to sample the *AEV_In* bus. It is important to note that aside from its simplicity, this stacked scheme also minimizes the possible leakage from the *AEV* capacitor, thus maintaining the data integrity till the readout. The readout of the *AEV* is performed through a separate source follower, implemented by M_{10} and M_{11}. This relatively simple pixel was implemented in layout using 0.18 μm CMOS technology, with 14 μm pitch and 40% fill factor (FF) (Figure 6).

Figure 6. Layout of **TYPE I** Pixel.

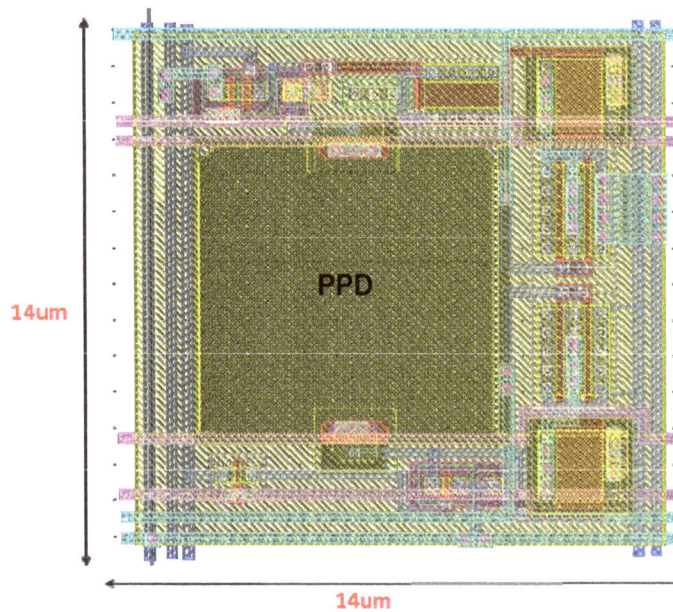

Figure 7. Phase 1 **TYPE I** pixel.

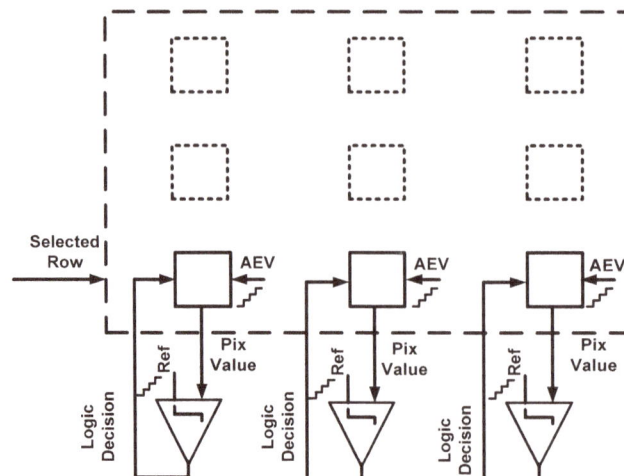

Figure 7 depicts Phase 1 for the **TYPE I** pixel. At each span, the pixel (*Pix value*) is being accessed by election of a specific row and is being compared to the reference signal *Ref*.

After Phase 1 is completed, Phase 2 begins. During this phase, each pixel integrates in accordance to the assigned *AEV*. After a certain integration slot (3) elapses, the *AEV*'s within each row are selected and compared with the *Ref*, which now reflects the reference values of integration times (Figure 8).

Figure 8. Phase 2 **TYPE I** pixel.

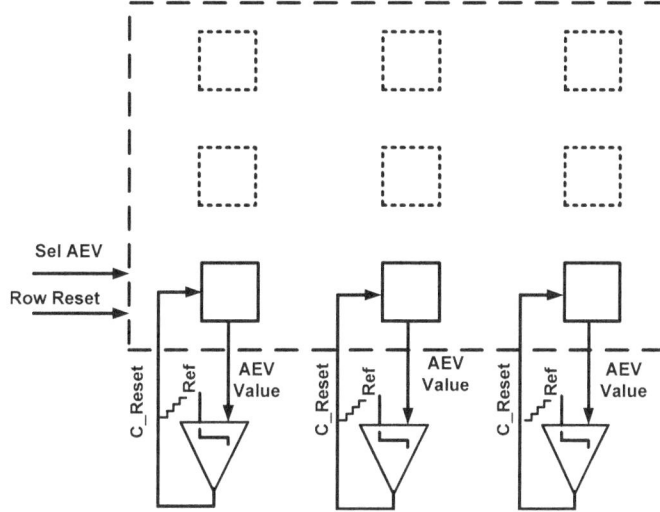

In our solution, the pixel is reset as long as its *AEV* is lower than the reference value.

In a **TYPE I** solution, the conditional resets are applied to each pixel row sequentially. The processing time, which takes to apply the conditional reset a single pixel row T_{row_reset}, is given by:

$$T_{row\_reset} = T_{AEV\_read} + T_{compare} \tag{4}$$

where T_{AEV_read} and $T_{compare}$ are the time to sample the *AEV* and to decide to reset or not the referred pixel, respectively. Since two saturation checks cannot overlap each other, the time it takes to complete a single one sets the minimal integration time $T_{min\_TYPE\ I}$:

$$T_{min\_TYPE\ I} = N_{rows} T_{row\_reset} \tag{5}$$

Given the minimal integration time, we can easily calculate the maximal possible DR extension factor (DRF) [8]. Through the performed simulations, we learned that the minimal integration time $T_{min\_TYPE\ I}$ is 112 μs. Assuming the maximal integration time is 30 ms, we have obtained a DR extension factor equal to 256 (48 dB). The expected intrinsic DR *i.e.*, before the extension is 60 dBs, consequently the total DR equals 108 dBs.

In the last phase of the frame (Figure 9), the accumulated data are accessed for the final ADC. By activating two separate select signals *Sel_Pix* and *Sel_AEV*, the pixel and the *AEV*, respectively, are digitized using the same comparators as in Phase 1

The **TYPE I** design was successfully tested in various post-layout simulations, which proved its feasibility. To reduce the length of this work, we present only the post-layout simulation results of the **TYPE II** design because it is much more complex. To conclude, CMOS sensor based upon the presented **TYPE I** pixel is capable of providing a DR of almost 5 decades operating at video frame rate and providing an image with a decent resolution.

Figure 9. Phase 3 **TYPE I** pixel.

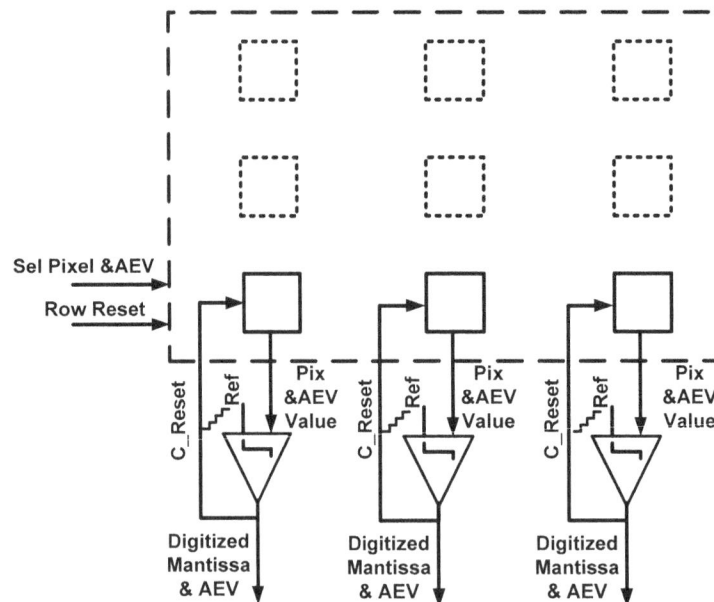

2.7. Ultra WDR (TYPE II) Pixel

The main difference between the Ultra WDR (**TYPE II**) pixel and the previous **TYPE I** solution lies in its accommodating a differential stage (D) within the pixel (Figure 10).

Figure 10. Schematic of **TYPE II** Pixel.

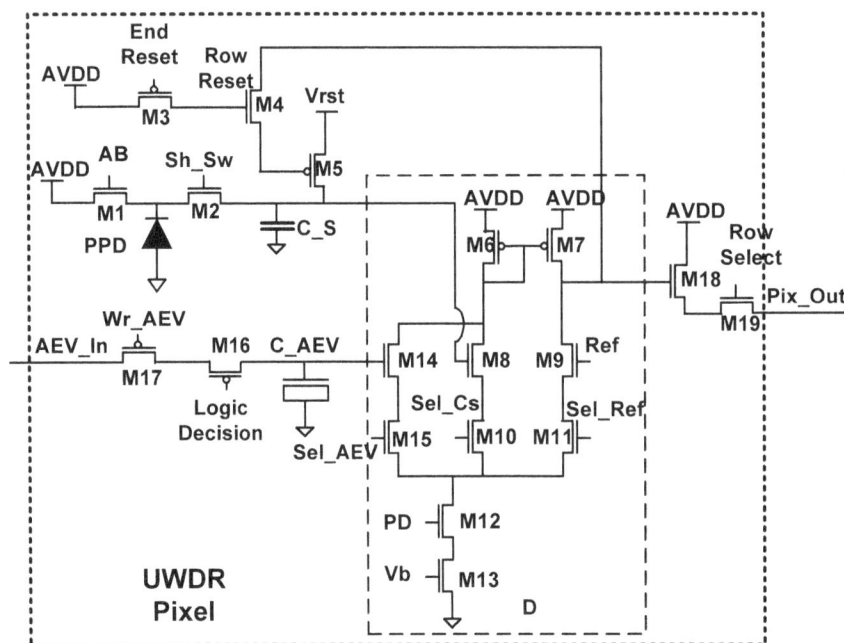

The inputs to this stage are: the pixel integration time represented by AEV (M_{14}); the photo-generated signal (M_8); and the global reference signal Ref (M_9). These inputs are sampled from

C_{AEV}, C_S, and *Ref*, respectively. By means of *Sel_AEV* (M_{15}) or *Sel_C$_s$* (M_{10}), the differential stage compares the *AEV* or the photo-generated signal, to the *Ref*. the transistor M_{11}, controlled by *Sel_Ref* signal, was added to ensure an optimal matching within the differential amplifier. The bias point is controlled by an analog signal V_b (M_{13}), which sets the current magnitude throughout the amplifier branches. To facilitate the power reduction, we inserted an additional transistor M_{12} in the series to the current source in order to shut down the entire stage, when no comparison is needed. The load of the *D* stage is implemented by two PMOS transistors: M_6 and M_7. The drain of the latter is connected to the drain of M_4, forming a self-reset structure.

The self-reset feature of the *D* amplifier enables the application of the conditional reset operation during a saturation check simultaneously to every pixel within the array rather than row by row. At each saturation check, the *Row Reset* (M_4) is raised globally, connecting the output of the *D* stage to the reset transistor M_5. Obviously, when the amplifier's output is low, the C_s capacitance is reset to V_{rst}, otherwise C_s remains without a change. To stop the reset cycle, regardless of the differential stage output, we use *End Reset* signal (M_3), which forces *AVDD* on the gate of the reset transistor M_5.

AEV assignments are done row by row as will be explained further. Before the assignment operation, the first *AEV* is asserted onto the *AEV_In* bus. Then, the signal *Wr_AEV* (M_{17}) selects the appropriate pixels' row. The *Logic Decision* signal (M_{16}), which is driven by the column-wise common source amplifiers, enables the corresponding *AEV* to be sampled by the C_{AEV} capacitor. The charge flow from the *PPD* is the same as in the **TYPE I** design throughout the whole frame.

At the end of the frame, both the *AEV* and the photo-generated charge are sampled nondestructively through the source follower (SF) (M_{18} and M_{19}) and converted by column-wise amplifiers, as will be explained later.

Figure 11. Layout of the **TYPE II** Pixel.

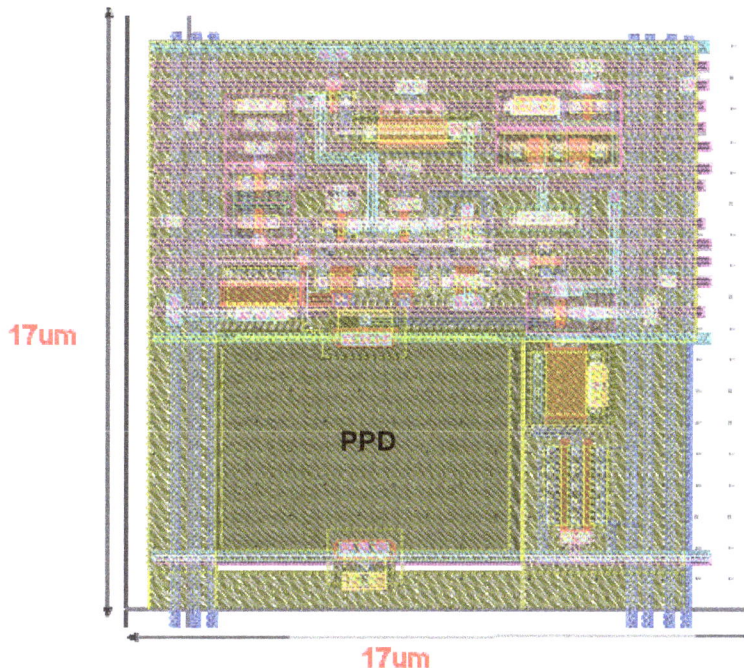

The **TYPE II** pixel was implemented in a 0.18 CMOS process (Figure 11). All the control signals were divided into two groups: 7 analog signals, and 10 logic signals. To reduce the coupling between the control lines, we used extensive orthogonal patterning, so that part of the signals was routed along the X axis, whereas another part was laid along the Y axis. The analog signals were implemented in Metal1 and placed along the Y axis to the left and to the right of the *PPD*. Such an arrangement reduced the coupling between the analog lines themselves and created a substantial separation from the rest logic signals, which were laid along the X axis. Moreover, all the signals, which were running horizontally, were implemented in Metal 3, which reduced the coupling between the pixel analog and logic lines even further. Due to the dense layout, we successfully implemented the described pixel with 17 μm pitch and 25% fill factor FF (Figure 11).

The three phases in the **TYPE II** solution are described in detail below. The *AEV* assignment that occurs in Phase 1 requires a high gain comparator. Since the gain of the in-pixel differential amplifier is not high enough, we added an external common source (*CS*) amplifier (Figure 12). In this way, we obtained the three stage high gain comparator: (1) *D*; (2) *SF*; (3) *CS*, respectively. The *Pix Value* is compared to *Ref* inside each pixel of the selected row. The final amplification is performed by *CS*, which drives the *Logic Decision* signal (Figure 12). This signal stamps the appropriate *AEV* value onto C_{AEV} capacitance inside each pixel. Figure 13 illustrates the flow of the *AEV* assignment to a pixel matrix of 2×2 at the first span. Each pixel within the matrix has its own coordinates: the first denotes the row and the second denotes the column. We performed the simulation as if each pixel within this matrix receives a different incoming signal. Therefore, we stimulated the pixels with different current sources, imitating different discharge rates. The *AEV* values were generated by a piecewise ascending ramp voltage ranging from 0.9 V up to 2.6 V, corresponding to 18 different values distanced by 0.1 V from each other. The first span occurs after 100ns. Then, the photo-generated charge is transferred to C_S simultaneously within the whole matrix and is compared row by row with the reference, thus causing the corresponding *AEV* to be written onto C_{AEV} capacitance (Figure 10). Capacitances C_{AEV}<0,0> and C_{AEV}<0,1>, found at the row <0>, are processed first, whereas row <1> capacitances are programmed after the completion of row <0> programming.

Figure 12. Phase 1 of **TYPE II** Pixel.

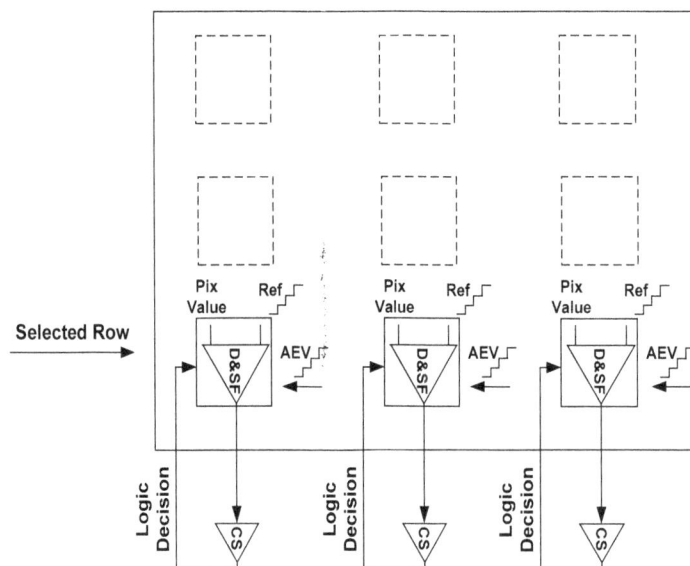

Figure 13. Post-Layout Simulation of Phase 1 of **TYPE II** Pixel.

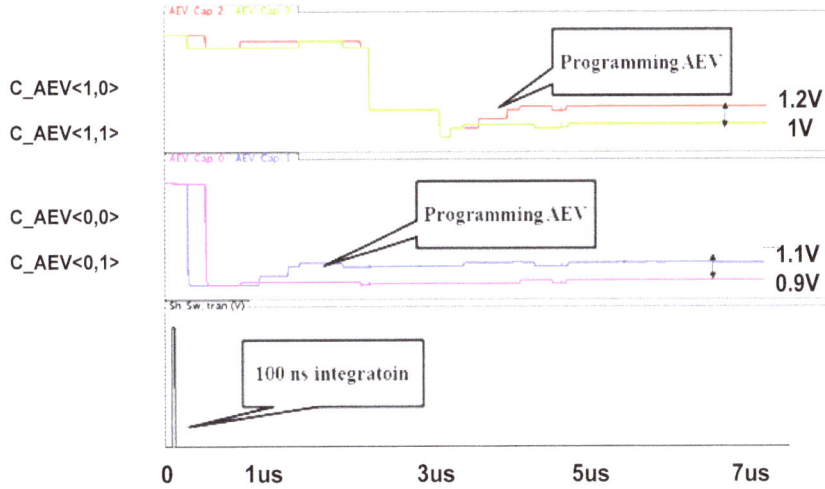

In the illustrated case, there are four different photo-generated signals, falling within four different segments; therefore every pixel is assigned a different *AEV* from 0.9 V to 1.2 V. The lowest *AEV* is assigned to the most illuminated pixel, *i.e.*, to $C_{AEV}<0,1>$, whereas the rest of the *AEV*'s: 1 V, 1.1 V, and 1.2 V are assigned to less illuminated pixels $C_{AEV}<1,1>$, $C_{AEV}<0,0>$, $C_{AEV}<1,0>$, respectively. It is important to note that the adjacent *AEV* values differ from each other by $2\Delta V_{min}$, which is 0.1 V in this case, and this difference stays intact after the *AEV*'s were written to C_{AEV}. In such a case, decoding the analog data at the end of the frame will be easy; otherwise the final pixel value will be erroneous due to a possible shift of the stored *AEV* values.

It is important to understand that the time of the first span is shorter than the minimal integration time, since, during all the spans, the available signal to be accumulated is lower than the pixel swing by $2\Delta V_{min}$. However, during the next phase the entire pixel swing is available, thus the real integration period will be somewhat higher. In the specific example we have discussed, the minimal integration time will be not 100 ns, but rather 112 ns.

Phase 2 in a **TYPE II** pixel is ultra-fast. Due to its self-reset ability, the reset decision inside each pixel is autonomously fed from the differential stage through M_4 to the gate of M_5 (Figure 10). Consequently, each pixel is reset independently upon the comparison of its own *AEV* with the global reference *Ref* (Figure 14). Therefore, the minimal integration time for this design is:

$$T_{min\_TYPE\,II} = T_{AEV\_read} + T_{compare} \tag{6}$$

Figure 14. Phase 2 of **TYPE II** Pixel.

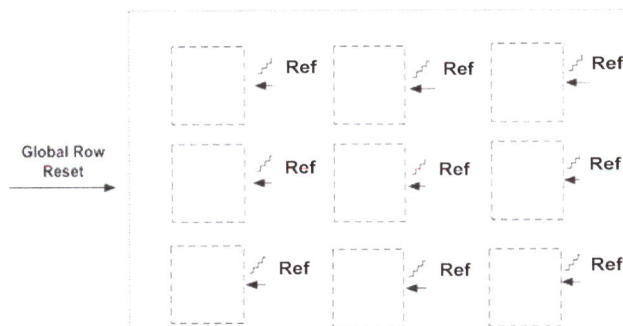

Equation (6), which describes the minimal pixel conditional reset time, sets the minimal integration time and, as such, defines the available DR extension. In this case, the minimal integration times can be set as low as 112ns, which enables the DR to be extended by over five decades.

Figure 15a,b illustrates the flow of Phase 2 on a matrix of 2 × 2 pixels. There are four different illuminated pixels, each with its own *AEV* in Phase 1. At Phase 2, every pixel is reset in accordance with its *AEV* (Figure 15a). The bottom pixel, C_S<0,1>, has the lowest *AEV* and therefore it is reset for five times; whereas the pixel C_S<1,0>with the highest *AEV* is reset only twice. A close–up of the typical reset sequence is presented in Figure 15b. First, the differential stage inside the pixel is activated by raising the *PD* signal; then, the *AEV* is compared to a reference by raising *Sel_AEV*. After this, the bias point of the amplifier stabilizes, the *Row Reset* signal is activated, which enables the decision to reach the reset transistor. In the specific case we have presented, the reset decision is positive and therefore the capacitance *Cs*<0,1> is charged up. The whole sequence lasts for 112 ns; thus, by assuming a typical frame time of 30 ms, we obtain DR extension of 109 dB [8]. Adding to this the intrinsic DR of 60 dBs (10 bits), we obtain overall DR of 170 dB approximately.

Figure 15. Post-Layout simulation of: (**a**) 4 different pixel signals during Phase 1; (**b**) a close up of a single self-reset cycle.

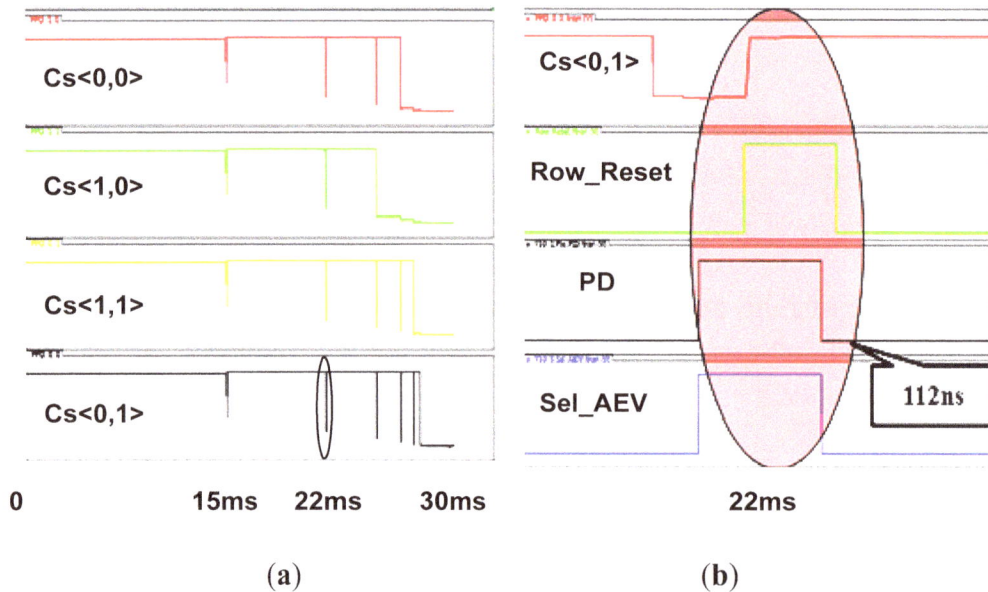

(a)   (b)

The final A/D conversion occurs at Phase 3. We use the same comparator configuration as depicted in Figure 12 to perform a single slope conversion to a photo-generated signal (*Mantissa*) and to an *AEV*. Figure 16 depicts the process of A/D conversion to the stored *AEV* and the pixel *Mantissa*. First, the pixel *AEV* is converted using a ramp with reduced ("coarse") resolution. It can easily be observed that the *Ref* signal spans the *AEV* domain in a reduced number of steps, which corresponds to the "coarse" conversion; whereas, the pixel *Mantissa* is spanned by the reference with a much higher ("fine") resolution. As a matter of fact, after the *Mantissa* is converted, the capacitance C_S storing it is being reset for the next frame. The capacitance holding the *AEV* is not necessary to be reset, since it will be assigned the new *AEV* in Phase 1 of the subsequent frame in any case.

Figure 16. Post-Layout Simulation of Phase 3 of **TYPE II** pixel.

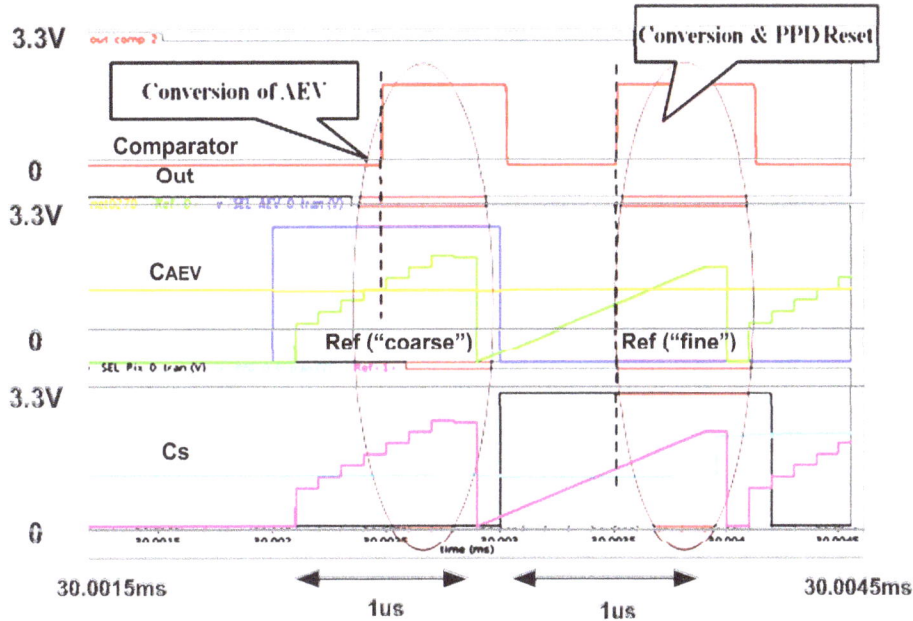

In conclusion, the relatively complex structure of the **TYPE II** pixel has enabled a very large DR extension and has substantially simplified the second and the third phases of the proposed algorithm. Figures 15 and 16 prove the feasibility of the **TYPE II** design and indicate that sensor based upon a **TYPE II** pixel can be successfully implemented in a silicon process.

2.8. A Power Profile of TYPE I and TYPE II Designs

Another important factor is the examination of the power performance of the **TYPE I** and **TYPE II** designs, since, if they were power hungry, it would be pointless using them. We made both of the designs optimized for low power using the following methods: the two stacked transistors scheme, and multiple power supplies. The result was that the leakage within the pixels was successfully minimized, and the power supplies could be lowered easily without degrading the anticipated sensor's performance.

In analyzing the **TYPE I** power per pixel (*PPX*), we have concluded that most of the power is consumed during the *AEV* assignment (Phase 1) and the A/D conversion (Phase 3), because these two operations require activating the comparators for a longer period of time than in Phase 2 (Figure 17a).

In **TYPE II**, the relative power contribution of Phase 2 is also the lowest, in comparison with the two other phases, as its overall effective duration is the shortest one (Figure 17b). Most of the power is consumed during Phase 1, which contains an increased number of spans and, thereby, in aggregate, lasts longer than Phase 3. Phase 3's contribution to *PPX*, is almost the same as in the **TYPE I** case due to similar flow of the data conversion.

Based on these simulations, we have concluded that the two presented designs consume power of 2.5 nW and 5.4 nW, respectively. Taking into account the anticipated extraordinary DR extension they provide, we have found that the power budget is low and definitely appropriate to the state-of-the art CMOS image sensors [25].

Figure 17. Power per Pixel Distribution of **TYPE I** and **TYPE II** Designs. (**a**) PPX TYPE I = 2.5 nW; (**b**) PPX TYPE I = 5.4 nW.

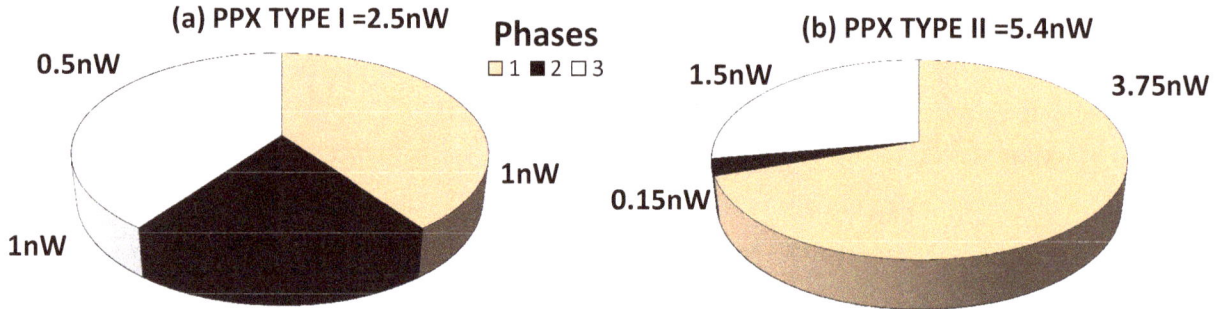

To summarize the anticipated performance of the proposed designs, we provide Table 1, where different state-of-the art WDR solutions are compared with respect to several key attributes such as fabrication technology, WDR technique (WDR T.), pixel size, fill factor (FF), DR, SNR, power per pixel (PPX), and frame rate (FR).

Table 1. Comparison between state-of-the art WDR image sensors.

Parameter	[11]	[5]	[12]	[14]	[21]	TYPE I	TYPE II
Technology	0.18 μm	0.09 μm	0.18 μm	0.35 μm	0.18 μm	0.18 μm	0.18 μm
WDR T.	Well Cap. Adjustment	Well Cap. Adjustment	Frequency	TFS	Multiple Capt.	Multiple Resets	Multiple Resets
Pixel Size	3 μm × 3 μm	5.86 μm × 5.86 μm	23 μm × 23 μm	81.5 μm × 76.5 μm	5.6 μm × 5.6 μm	14 μm × 14 μm	17 μm × 17 μm
FF	-	-	25%	2%	45%	40%	25%
DR	100 dB	83 dB	130 dB	100 dB	99d B	108 dB	170 dB
SNR	48 dB[*]	48 dB[*]	-	-	-	48 dB[*]	48 dB[*]
PPX	-	400 nW[**]	250 nW	≤6.4 μW[**]	10 nW[**]	2.5 nW	5.4 nW
FR	-	30	-	-	15	33	33

([*]): The SNR was assessed based upon the well capacity of the photo-diode; ([**]): The PPX was calculated by normalization of total power by the number of pixels within the array.

From Table 1, we can understand that the pixel sizes of pixels presented in our work are mediocre. They are substantially larger than those associated with well capacity adjustment and multiple captures, but much smaller than frequency and TFS based sensors. Important to note that both of our designs maintain a decent FF relatively other listed solutions due to area effective layout. The DR of proposed herein designs emphasizes the extraordinary ability of multiple resets algorithm to extend the pixel dynamic range. **TYPE I** provides remarkable DR, whilst **TYPE II** brings it to extreme values. Moreover, such DR is obtained, maintaining excellent SNR (see Appendix B). Not less important to note that both of proposed designs operate at video frame rate and present the best power performance.

3. A Comparative Analysis of Multiple Captures & Multiple Resets Algorithms

When a new WDR solution is proposed, it is always interesting to compare it to the existing ones and to realize the added value the new solution contributes. In the present discussion, we perform a comparison between the two proposed solutions to the following algorithms: the multiple captures [15] and the former multiple resets solutions [24]. We chose to compare the number of operations needed to obtain a single digitized WDR pixel value; because, in our opinion, the most significant contribution of **TYPE I** and **TYPE II** solutions is their effectiveness and their ability to substantially reduce the processing time.

First, the multiple captures algorithm must be considered. In this solution, the pixel integrates for several periods of time, regardless of the intensity of the incoming light. After each integration, the intermediate pixel value is synthesized out of both the newly generated and previously stored samples, respectively. At this stage we assume that all the captures are performed in a rolling shutter mode to comply with the minimal memory requirements for this solution. In case multiple captures are performed in a global shutter mode, the memory requirements will be much higher. For a rolling shutter operation mode, the number of memory bits for each pixel equals its digital resolution: $N_{resolution}$. The total number of operations that are required to obtain a final digitized result for a single pixel after $N_{captures}$ exposures is:

$$N_{op\_captures\_tot} = \left(N_{captures} - 1 \right)\left(1_{convert} + N_{resolution}\left(1_{read} + 1_{write} \right) \right) \tag{7}$$

where $1_{convert}$, 1_{read}, and 1_{write} stand for ADC, memory read, and write cycles, respectively. Please note, that in this case there are two captures only: the memory actually becomes a single row register where the first sample is kept, in order to be compared with the second.

On the other hand, both the previous and newly proposed multiple resets solutions are applied on a global shutter sensor to obtain the maximal possible number of operations. According to the snapshot mode, the number of memory bits for a single pixel equals Rep_{EXP}, which is the representation of the exponent bits N_{EXP}. Obviously, in case of linear representation, Rep_{EXP} equals N_{EXP}. If logarithmic representation is applied, the Rep_{EXP} is reduced to $\log_2 N_{EXP} + 1$ [27]. For example, if there are 6 exponent bits, then in the first case, Rep_{EXP} is 6, whereas in the second it is 4. The overall number of operations in previous solutions is:

$$N_{op\_resets\_tot\_prev} = N_{EXP}\left(1_{read} + 1_{convert} + 1_{write} \right) + 1_{read} Rep_{EXP} + 1_{convert} \tag{8}$$

The first term in (8), relates to the sequential generation of DR bits, which consists of memory read, pixel conversion by the comparator, and memory write cycles. The second term relates to the digital readout of the memory contents, and the last term denotes the A/D conversion of the pixel *Mantissa* at the end of the frame.

In the newly presented **TYPE I** and **TYPE II** solutions, the total number of operations coincides with:

$$N_{op\_resets\_tot\_new} = 1_{convert}\left(l + N_{EXP} + 2 \right) \tag{9}$$

The term *l* stands for the number of spans performed during Phase 1; N_{EXP} is for Phase 2, and the last term of 2 denotes the number of conversions during Phase 3.

Next, we will demonstrate the meaning of Equations (7–9) by substituting the possible values for each parameter. We have presented the results in Table 2, where each row is allocated for a different $N_{captures}$ and N_{EXP}.

For the previous solutions of multiple resets, we chose the logarithmical representation as in [27], because it minimizes the required external memory size. We also have assumed that the number of the exponent bits N_{EXP} equals the number of captures: $N_{captures}$. This assumption not only simplifies the calculations, but it also ensures that the SNR dips, which exist throughout the extended DR in all of the solutions, are equal.

The advantage of the multiple captures algorithm is that most of it is implemented in a digital domain and the pixel structure is rather simple. The clear disadvantage is that this solution requires an extensive processing to obtain the final result. From Equation (7), it is shown that the number of operations increases linearly with the number of captures. This fact is clearly demonstrated by Table 2, where the multiple captures solution obviously possesses the highest number of operations per pixel. It is possible, though, to limit the number of captures to two or three; however this will cause large SNR dips at the boundary light intensities [8]. Circumventing these SNR artifacts by adding more integration slots will certainly slow the sensor operation speed. Consequently, the multiple captures algorithm is best suited to photograph slow changing scenes with a very high spatial resolution.

Table 2. The Comparison of a number of operations for "Multiple Captures", "Multiple Resets Previous", and "Multiple Resets New" algorithms.

$N_{captures} = N_{EXP}$	$N_{EXP} = N_{captures}$		
	Multiple Captures [15]	**Multiple Resets Previous [27]**	**Multiple Resets New (This Work)**
2	21	9	5
3	42	13	6
4	63	16	7
6	105	23	10
18	357	61	25

In the multiple resets algorithm, although the pixel structure and the pixel control are more complex than in the first case, the overall processing is much faster, since the pixel adjusts the integration time autonomously during the frame. Thus, there is no need for an extensive post-integration processing—only to combine the data of the *Mantissa* and the *EXP* bits. From Table 2, we realize that even the previous configurations of multiple resets have a reduced number of operations, relative to the multiple captures algorithm. The intuitive explanation of this is found within a sequential processing of the pixel information, which allows reading only the last produced bit from the memory before generating the subsequent one. The logarithmical representation of the *EXP* bits, used in the previous designs, decreases the operations number even further. Therefore, the previous solutions of multiple resets were perfectly suited to capture fast changing WDR scenes with a decent resolution.

The two solutions, proposed within the current study, bring the number of operations to its minimum, since they completely eliminate the operations related to a digital memory unit. This feature enables the proposed solutions to be the fastest among the three, which have been analyzed in this discussion. The clear disadvantage of the pixel using *AEV* is a relatively complex analog signal processing inside each pixel, which is especially pronounced in **TYPE II** design. Therefore, the newly proposed solutions can be candidates for implementation for capturing a very fast changing scene with wide or ultra-wide dynamic range.

4. Summary

We have presented two designs: **TYPE I** and **TYPE II** featuring the "ranging" and the *AEV* concepts. We have shown that it is possible to perform effective encoding of the DR bits using analog voltage, which can be easily stored within the pixel and processed further through the readout chain. Moreover, we have condensed the *AEV* assignment to a single phase in a process called "ranging", which has led to another improvement: a substantial simplification of the conditional reset. In the past, each cycle consisted of three or even more different operations; now the number of operations has been nearly halved. The *AEV* solution has also granted a possibility to save chip area due to reuse of high gain comparators as a main block in a single slope ADC. The post-layout simulations and power calculations included in our study have shown that both the **TYPE I** and the **TYPE II** designs are feasible and can be successfully implemented in a CMOS process. The presented designs have reached a remarkable DR of up to 170 dB, while only consuming power of up to 5.4 nW. A further comparison between this work and the multiple captures and the former multiple resets solutions has clearly proven that the *AEV* concept effectively reduces the number of operations required to produce a pixel value and, as such, can be considered as a key to design a fast, low-power and ultra WDR CMOS sensor.

Appendix A

Division of the Pixel Swing to Segments

Since our algorithm is based upon integration times ordered in geometrical progression, we divide the pixel swing S_{pix} into geometrically ordered segments as well: $S_1, S_2, ..., S_n$ as depicted in Figure 18.

During the frame, the pixel discharges from the *Reset level* towards the noise floor. The *Reset level* and the lower bound of pixel signal define the pixel swing S_{pix}. The first segment S_1 equals $2\Delta V_{min}$. Consequently, the subsequent segments will be equal to $4\Delta V_{min}$, $8\Delta V_{min}$, and so on, till the last segment S_n, (Figure 18). In the general form, the values of these segments are given by:

$$S_i = 2^{i-1} \cdot \left(2\Delta V_{min}\right) \leq S_{pix} - 2\Delta V_{min} \quad 1 \leq i \leq n \tag{10}$$

It is important to note that the last segment S_n, is distanced from the pixel swing S_{pix} by $2\Delta V_{min}$. This means that signals exceeding S_n will not be converted. The purpose of this restriction will be explained later on.

That the number of *AEV* values that can be assigned in a single span equals the number of segments, which are given from (10) can be easily understood. For example, as shown in (Figure 18) if the pixel S_n equals $16\Delta V_{min}$, there will be four different segments; thus, four values of *AEVs* can be assigned during a single conversion.

The ramp signal *Ref* spans the pixel signal. This signal has to comply with the following requirements: (1) the signals that reach the lower bound of S_{pix} will not be converted; and (2) signals reaching the edge of the last segment, S_4 in this case, will be converted. In this way, the ramp starts on its way in the middle of the restricted region marked in red (Figure 19).

Figure 18. Division of the Pixel Swing to Segments.

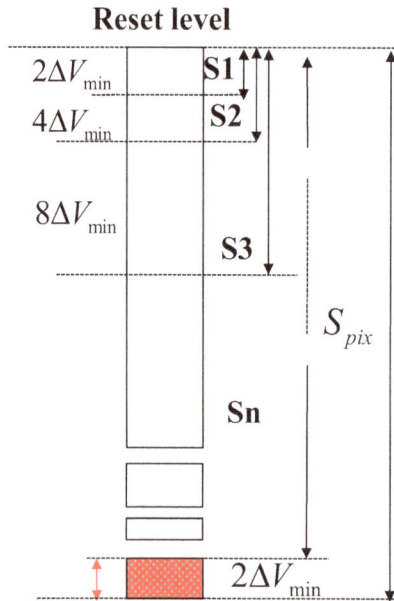

Figure 19. Conversion of a pixel swing composed out of four segments.

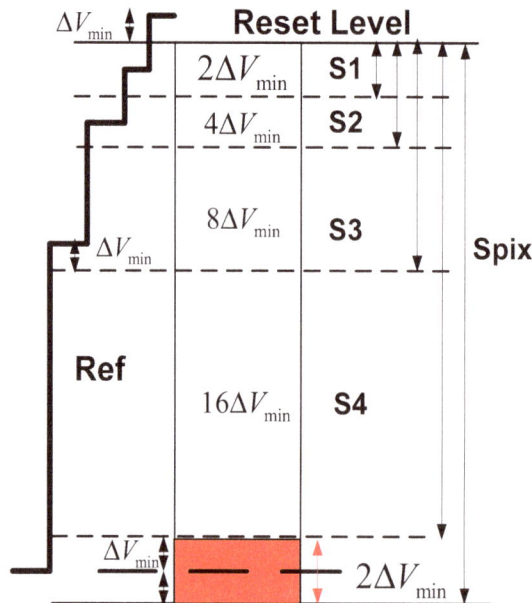

From there, it climbs in geometrically decreasing steps, gradually covering all the pixel segments. To insure that all the segments are duly covered, the ramp *Ref* surpasses each one by ΔV_{min}. An additional positive consequence of such ramp configuration is that the minimal SNR during each conversion is set by ΔV_{min} (see Appendix B).

Special attention needs to be paid to the synchronization between the change of *AEV* and the ramp steps (Figure 20). Please note that *AEV* 2 becomes valid before the *Ref* continues to climb and it stays constant for the $t_{program}$ interval, during which the *AEV* 2 value is being written onto the C_{AEV}.

Figure 20. *AEV* change in correlation with the comparator reference.

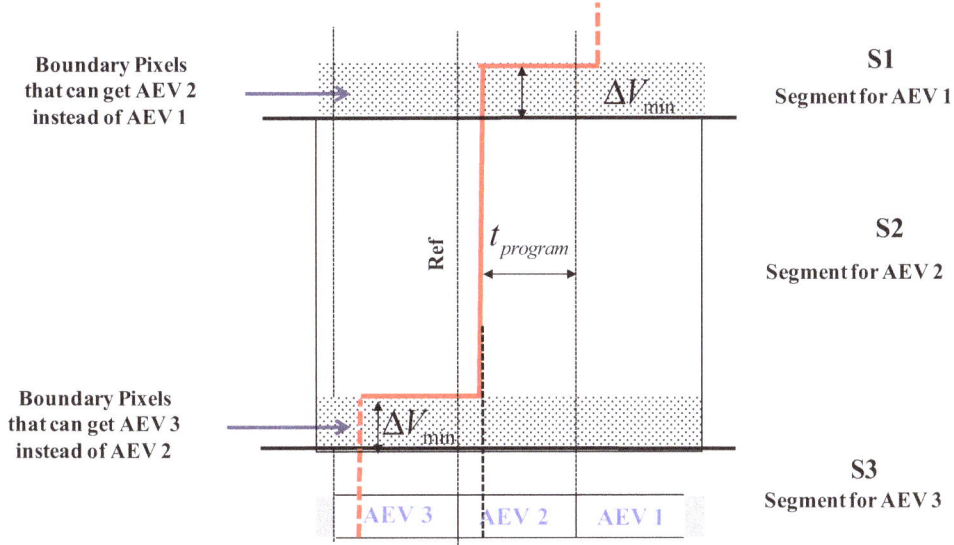

Another essential aspect is the assignment of the *AEV* values to the pixels, the signals of which lie at the boundaries of the adjacent segments. Such *Boundary Pixels* can receive the *AEV* value corresponding to the previous segment. For example, the ramp *Ref* finalizes spanning the pixel signals that should receive *AEV* 2 within the adjacent S_1 segment. In such a situation, some pixels, found at the lower boundary of a S_1 segment can mistakenly receive *AEV* 2 instead of *AEV* 1. As a result, the pixels' saturation is prevented, but this is at the expense of the reduction of half of their effective integration time, which leads to a SNR drop of 6 dBs.

Up to this point, we have explained the aspects of a single pixel span, but what happens if we were to perform additional spans to complete the desired number of *EXP* bits N_{EXP}? Then, the number of spans l is increased to (11). For example, if we suppose there are 18 bits to be encoded into 18 *AEV* values and in each span 4 bits are encoded, then 5 scans are needed to complete Phase 1.

$$ l = \left\lceil \frac{N_{EXP}}{n} \right\rceil \tag{11} $$

where n is the number of segments. Since there might be several spans, how are we to ensure that *AEV*s, which were assigned in the preceding spans, are not overridden in the subsequent ones? The answer is to distance the last segment by $2\Delta V_{min}$ from the pixel lower boundary. If we look for the reason the *AEV*'s were not overridden, we learn that there are two spans necessary to encode eight different light intensities "1–8" (Figure 21a,b). After the first span, the four most significant intensities "5–8" are covered and assigned their *AEV*'s (Figure 21a). The weaker four intensities ("1–4") are assigned *AEV*'s, which are not correct and therefore should be overwritten in the upcoming second span. During the second span, all the pixels will be addressed once again, regardless if they have already received their *AEV*'s or not. The second span is distanced from the first, so that the signal generated by "5–8" exceeds the last segment by $2\Delta V_{min}$. Therefore, these pixels do not receive new *AEV*'s. On the other hand, "1–4" intensities generate signals, which now fall within the different segments and thereby are assigned the corresponding *AEV*'s (Figure 21b). Hence, by the reduction of the possible conversion range, we can differentiate between the already encoded pixels and those which should be encoded in the future spans.

Figure 21. Distribution of the different light intensities by segments (**a**) first span; (**b**) second span.

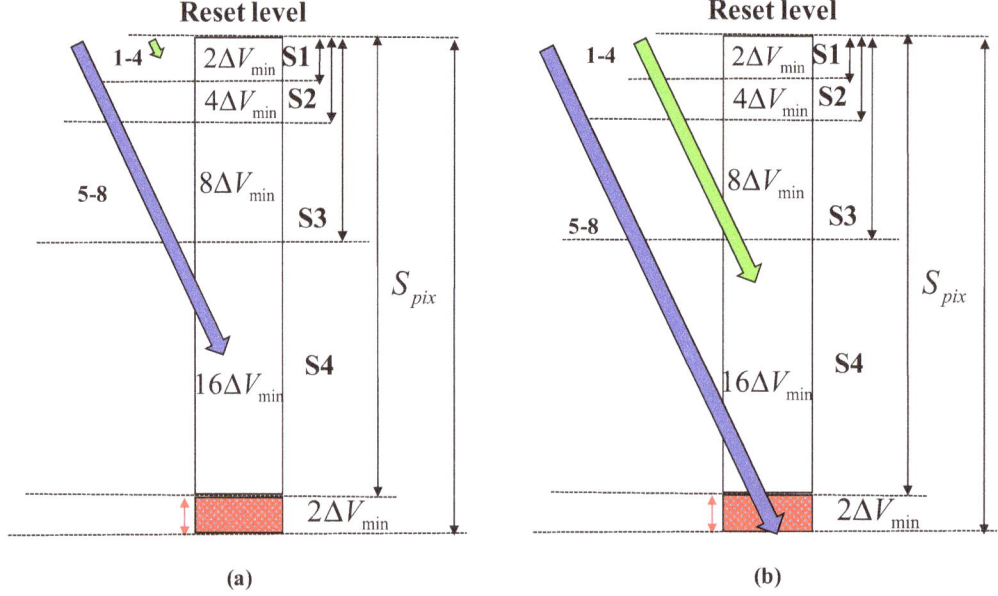

(a)   (b)

All the spans are spread non-uniformly in time. The occurrence of each span t_k is set upon the most significant *EXP* bit that can be encoded in the k-th span:

$$t_k = \frac{T_{int}}{2^{M_k}} \cdot \frac{S_{pix} - 2\Delta V_{min}}{S_{pix}} \quad k = 1, 2, ..., l \tag{12}$$

where M_k is the MSB that can be encoded in the k-th span. For example, if there are 18 *EXP* bits to be encoded during 5 spans, then $M_1 = 18$; $M_2 = 14$; $M_3 = 10$; $M_4 = 6$; $M_5 = 2$, respectively. Furthermore, assuming that $S_{pix} = 1$ V; $2\Delta V_{min} = 0.112$ V; $T_{int} = 30$ ms, the span times t_1 till t_5 are: 100 ns; 1.65 µs, 26.4 µs, 422 µs, and 6.75 ms, respectively.

Appendix B

SNR Considerations

SNR is one of the key properties of every sensor. Every designer must be aware how much noise each specific circuitry architecture generates. Herein, we discuss three noise sources shot noise, flicker noise, and the thermal noise. We consider the noise of the two types of the sensors separately. The performed analysis shows that in poorly illuminated scenes, the thermal noise governs the noise floor for both of the sensors, whereas the shot noise of photo-diode sets the noise limit in highly illuminated scenes. We refer to noise induced to *AEV* capacitance as well, and show that our design considerations provide an excellent signal to noise ratio of the stored data throughout the whole frame.

TYPE I

We assume, for our convenience that all noise sources are uncorrelated, therefore we can write the overall SNR as:

$$SNR = 20\log \frac{i_{ph}t_{int}}{\sqrt{V_{n\_KTC}^2 + V_{n\_shot}^2 + V_{n\_read}^2}} \tag{13}$$

where V_{n_KTC}, V_{n_shot}, V_{n_read} are the voltage variances of the thermally generated KTC noise, shot noise, and read noise, respectively. i_{ph} denotes the photo-generated current and t_{int} denotes the time it is integrated within the photo-diode.

The noise analysis for this sensor is straightforward, due to its conventional structure. At the beginning of the frame, the in-pixel sense capacitance C_S (Figure 5) is being reset, which causes well-known KTC noise. Since it is "hard reset" [28], performed through PMOS, the overall inflicted noise equals:

$$V_{n\_KTC} = \sqrt{\frac{k_B T}{C_S}} \tag{14}$$

where k_B and T are the Boltzman's constant and junction temperature in degrees of Kelvin, respectively. According to our assessments, the resulted noise equals to 643 µV. Note, that since at the end of the frame, we do not need to convert the reset signal as in other conventional designs, this noise is not doubled. In case, there are multiple resets applied, this noise contribution grows by the square root of the reset times.

Shot noise within the photodiode is inflicted by both the dark current i_{dc} and the incoming optical signal i_{ph} [28]. The overall noise contribution is given by:

$$V_{n\_shot} = \sqrt{\frac{q(i_{ph} + i_{dc})t_{int}}{C_{PD}^2}} \tag{15}$$

In case the accumulated charge is maximal, *i.e.*, it equals the pixel swing S_{pix}, we get the following noise:

$$V_{n\_shot\_max} = \sqrt{\frac{qS_{pix}}{C_{PD}}}, \quad SNR_{max} = 20\log \frac{S_{pix}}{V_{n\_shot\_max}} = 20\log \sqrt{\frac{C_{PD}S_{pix}}{q}} \tag{16}$$

According to our assessments this amount of noise can reach 4 mV. In such a case the maximal SNR according to (16) equals 48 dB.

The readout noise in this sensor is mainly generated by the in-pixel source follower amplifiers, through which the readouts of *AEV* and the pixel are performed. It is easy to show that the noise source follower amplifier generates [29] is given by:

$$V_{n\_readout\_SF} = \sqrt{\frac{2k_B T}{3C_{Pix\_Out}}} \quad or \quad \sqrt{\frac{2k_B T}{3C_{AEV\_Out}}} \tag{17}$$

where C_{Pix_Out}, C_{AEV_Out}, are the sample capacitances for pixel and the *AEV* readout, respectively. One should note that since the gain of source follower never exceeds unity, the load capacitances should be large enough to reduce the readout noise to desirable level. In our case, the sample capacitances were designed to be no less than 100 fF, which bounded the readout noise to 200 μV, approximately.

It must be noted that there are additional noise sources within the source follower amplifier such as: flicker and telegraph noise [28]. The power of these noise sources is strongly depends on the process quality. With the recent fabrication technology advances, the total contribution of theses noise sources has been reduced to couple of electrons rms, which in our case equal to couple of tens micro-volts.

Additional noise sources, which have to be taken into consideration during the readout, is the thermal noise generated by the column-wise high gain comparator (Figure 7). The comparator consists of two stages: differential and common source. It is easy to show that the input referred thermal noise voltage variance is given by [29]:

$$V_{n\_readout\_Diff} = \sqrt{\frac{4k_B T}{3C_{CS\_in}G_{Diff}}} \tag{17}$$

where C_{CS_in} is the common source input capacitance and G_{Diff} is the differential stage gain. Correspondingly, the noise generated by the common source is:

$$V_{n\_readout\_Diff} = \sqrt{\frac{2k_B T}{3C_{CS\_Out}G_{Diff}G_{CS}}} \tag{18}$$

Substituting the corresponding values into (18), (19), we assessed that the input referred noise coming from the comparator is 55 μV, which is substantially lower than the one induced by the source follower.

The in-pixel dynamic capacitance storing *AEV* value also accommodates a certain amount of noise. Apparently, it is the KTC noise, which is injected onto C_{AEV} (Figure 5) the moment the *Logic Decision* signal disconnects it from the *AEV_In* bus. The amount of noise can be calculated by (14) by substituting C_{AEV} value into that formula. Since C_{AEV} is almost the same as C_S the noise equals 643 μV as well. Taking into account that the minimal difference between the *AEV* value and the reference signal *Ref* (Figure 19) is ΔV_{min}, we get the following minimal SNR:

$$SNR_{min} = 20\log_{10}\sqrt{\frac{C_{AEV}\Delta V_{min}}{k_B T}} \tag{19}$$

For example, if C_{AEV} is 10 fF and ΔV_{min} is 50 mV, we can observe that the minimal SNR is 37 dB, which is enough for obtaining 4–5 bits resolution during each span.

To conclude, **TYPE I** sensor reaches 48 dB SNR, when exposed to high light intensities. The dominant noise at low light is the KTC, whereas at high end it is the shot noise, which limits the sensor performance.

TYPE II

The SNR in this sensor is also given by (13). The first two terms of noise, namely V_{n_KTC}, V_{n_shot}, are similar to the **TYPE I** sensor. The readout noise, however, is much lower due to accommodation of the differential stage inside the pixel. In this configuration, the noise contributions of source follower, common source become reduced by the gain of the in-pixel differential stage. Thus, employing Equations (17)–(19), we can assess that the input referred readout noise does not exceed 55–60 µV. Obviously, the noise floor of **TYPE II** varies almost the same as of **TYPE I** throughout the whole illumination range.

The noise associated with *AEV* is governed by the KTC noise injected during Phase 1 and therefore equals the one for **TYPE I** sensor.

References

1. Mizobuchi, K.; Adachi, S.; Tejada, J.; Oshikubo, H.; Akahane, N.; Sugawa, S. A low-noise wide dynamic range CMOS image sensor with low and high temperatures resistance. *Proc. SPIE* **2008**, *6816*, 681604:1–681604:8.

2. Stevens, E.; Rochester, N.Y. Low-Crosstalk and Low-Dark-Current CMOS Image-Sensor Technology Using a Hole-Based Detector. In *Proceedings of the ISSCC—Image Sensors and Technology*, San-Francisco, CA, USA, 3–7 February 2008; pp. 60–62.

3. Shcherback, I.; Belenky, A.; Yadid-Pecht, O. Empirical dark current modeling for complementary metal oxide semiconductor active pixel sensor. *Opt. Eng.* **2002**, *41*, 1216–1219.

4. Kavadias, S.; Dierickx, B.; Scheffer, D.; Alaerts, A.; Uwaerts, D.; Bogaerts, J. A logarithmic response CMOS image sensor with on-chip calibration. *IEEE J. Solid-State Circuits* **2000**, *35*, 1146–1152.

5. Sakakibara, M.; Oike, Y.; Takatsuka, T.; Kato, A.; Honda, K.; Taura, T.; Machida, T.; Okuno, J.; Ando, A.; Fukuro, T.; *et al.* An 83dB-Dynamic-Range Single-Exposure Global-Shutter CMOS Image Sensor with In-Pixel Dual Storage. In *Proceedings of the IEEE International Solid State Circuits Conference Digest Technical Papers (ISSCC)*, San Francisco, CA, USA, 19–23 February 2012; pp. 380–382.

6. Seo, M.W.; Suh, S.; Iida, T.; Watanabe, H.; Takasawa, T.; Akahori, T.; Isobe, K.; Watanabe, T.; Itoh, S.; Kawahito, S. An 80 μV_{rms}-Temporal-Noise 82 dB-Dynamic-Range CMOS Image Sensor with a 13-to-19b Variable-Resolution Column-Parallel Folding-Integration/Cyclic ADC. In *Proceedings of the IEEE International Solid-State Circuits Conference Digest of Technical Papers (ISSCC)*, San Francisco, CA, USA, 20–24 February 2011; pp. 400–402.

7. Lu, J.H.; Inerowicz, M.; Joo, S.; Kwon, J.K.; Jung, B. A low-power, wide-dynamic-range semi-digital universal sensor readout circuit using pulse-width modulation. *Sens. J. IEEE* **2011**, *11*, 1134–1144.

8. Spivak, A.; Belenky, A.; Fish, A.; Yadid-Pecht, O. Wide-dynamic-range CMOS image sensors—Comparative performance analysis. *Electron. Devices IEEE Trans.* **2009**, *56*, 2446–2461.

9. Storm, G.; Henderson, R.; Hurwitz, J.E.D.; Renshaw, D.; Findlater, K.; Purcell, M. Extended dynamic range from a combined linear–logarithmic CMOS image sensor. *IEEE J. Solid State Circuits* **2006**, *41*, 2095–2106.

10. Akahane, N.; Sugawa, S.; Adachi, S.; Mori, K.; Ishiuchi, T.; Mizobuchi, K. A sensitivity and linearity improvement of a 100-dB dynamic range CMOS image sensor using a lateral overflow integration capacitor. *IEEE J. Solid State Circuits* **2006**, *41*, 851–858.

11. Sakai, S.; Tashiro, Y.; Kawada, S.; Kuroda, R.; Akahane, N.; Mizobuchi, K.; Sugawa, S. Pixel scaling in complementary metal oxide silicon image sensor with lateral overflow integration capacitor. *Jpn. J. Appl. Phys.* **2010**, *49*, 1–6.

12. Wang, X.; Wong, W.; Hornsey, R. A high dynamic range CMOS image sensor with in-pixel light-to-frequency conversion. *IEEE Trans. Electron. Devices* **2006**, *53*, 2988–2992.

13. Kitchen, A.; Bermak, A.; Bouzerdoum, A. PWM digital pixel sensor based on asynchronous self-resetting scheme. *IEEE Electron. Device Lett.* **2004**, *25*, 471–473.

14. Leñero-Bardallo, J.A.; Serrano-Gotarredona, T.; Linares-Barranco, B. A five-decade dynamic-range ambient-light-independent calibrated signed-spatial-contrast AER retina with 0.1-ms latency and optional time-to-first-spike mode. *Circuits Syst. I IEEE Trans.* **2010**, *57*, 2632–2643.

15. Mase, M.; Kawahito, S.; Sasaki, M.; Wakamori, Y.; Furuta, M. A wide dynamic range CMOS image sensor with multiple exposure-time signal outputs and 12-bit column-parallel cyclic A/D converters. *Solid State Circuits IEEE J.* **2005**, *40*, 2787–2795.

16. Dattner, Y.; Yadid-Pecht, O. High and low light CMOS imager employing wide dynamic range expansion and low noise readout. *Sens. J. IEEE* **2012**, *12*, 2172–2179.

17. Sandhu, T.S.; Pecht, O.Y. New memory architecture for rolling shutter wide dynamic range CMOS imagers. *Sens. J. IEEE* **2012**, *12*, 767–772.

18. Woo, D.H.; Nam, I.K.; Lee, H.C. Smart reset control for wide-dynamic-range LWIR FPAs. *Sens. J. IEEE* **2011**, *11*, 131–136.

19. Stoppa, D.; Vatteroni, M.; Covi, D.; Baschirotto, A.; Sartori, A.; Simoni, A. A 120-dB dynamic range CMOS image sensor with programmable power responsivity. *IEEE J. Solid State Circuits* **2007**, *42*, 1555–1563.

20. Yasutomi, K.; Itoh, S.; Kawahito, S. A two-stage charge transfer active pixel CMOS image sensor with low-noise global shuttering and a dual-shuttering mode. *Electron. Devices IEEE Trans.* **2011**, *58*, 740–747.

21. Choi, J.; Park, S.; Cho, J.; Yoon, E. A 1.36μW adaptive CMOS Image Sensor with Reconfigurable Modes of Operation from Available Energy/Illumination for Distributed Wireless Sensor Network. In *Proceedings of the 2012 IEEE International Solid-State Circuits Conference Digest of Technical Papers (ISSCC)*, San Francisco, CA, USA, 19–23 February 2012; pp. 112–114.

22. Han, S.W.; Kim, S.J.; Choi, J.H.; Kim, C.K.; Yoon, E. A High Dynamic Range CMOS Image Sensor with In-Pixel Floating-Node Analog Memory for Pixel Level Integration Time Control. In *Proceedings of the Symposium on VLSI Circuits Digest of Technical Papers*, Hilton Hawaiian Village, Honolulu, HI, USA, 13–15 June 2006; pp. 25–26.

23. Snoeij, M.F.; Theuwissen, A.J.P.; Makinwa, K.A.A.; Huijsing, J.H. Multiple-ramp column-parallel ADC architectures for CMOS image sensors. *Solid State Circuits IEEE J.* **2007**, *42*, 2968–2977.

24. Spivak, A.; Teman, A.; Belenky, A.; Yadid-Pecht, O.; Fish, A. Power-performance tradeoffs in wide dynamic range image sensors with multiple reset approach. *J. Low Power Electron. Appl.* **2011**, *1*, 59–76.

25. Spivak, A.; Teman, A.; Belenky, A.; Yadid-Pecht, O.; Fish, A. Low-voltage 96 dB snapshot CMOS image sensor with 4.5 nW power dissipation per pixel. *J. Sens.* **2012**, *12*, 10067–10085.

26. Yadid-Pecht, O.; Fossum, E.R. Image sensor with ultra-high-linear dynamic range utilizing dual output CMOS active pixel sensors. *IEEE Trans. Electron. Devices* **1997**, *44*, 1721–1724.

27. Belenky, A.; Fish, A.; Spivak, A.; Yadid-Pecht, O. Global shutter CMOS image sensor with wide dynamic range. *IEEE Trans. Circuits Syst. II Exp. Briefs* **2007**, *54*, 1032–1036.

28. Tian, H.; Fowler, B.; Gamal, A.E. Analysis of temporal noise in CMOS photodiode active pixel sensor. *Solid State Circuits IEEE J.* **2001**, *36*, 92–101.

29. Razavi, B. *Design of Analog CMOS Integrated Circuits*; McGraw-Hill Higher Education: New York, NY, USA, 2001; pp. 209–240.

30. Janesick, J.R. *Photon Transfer DN to λ*; SPIE Library: Bellingham, WA, USA, 2007; pp. 163–164.

Low Power Dendritic Computation for Wordspotting

Suma George *, Jennifer Hasler, Scott Koziol, Stephen Nease and Shubha Ramakrishnan

Georgia Institute of Technology, Atlanta 30363, GA, USA;
E-Mails: jennifer.hasler@ece.gatech.edu (J.H.); scott.m.koziol@gmail.com (S.K.);
stephen.h.nease@gmail.com (S.N.); shubha@gatech.edu(S.R.)

* Author to whom correspondence should be addressed; E-Mail: suma.george@gatech.edu

Abstract: In this paper, we demonstrate how a network of dendrites can be used to build the state decoding block of a wordspotter similar to a Hidden Markov Model (HMM) classifier structure. We present simulation and experimental data for a single line dendrite and also experimental results for a dendrite-based classifier structure. This work builds on previously demonstrated building blocks of a neural network: the channel, synapses and dendrites using CMOS circuits. These structures can be used for speech and pattern recognition. The computational efficiency of such a system is >10 MMACs/μW as compared to Digital Systems which perform 10 MMACs/mW.

Keywords: computational modeling; hidden markov models; neuromorphic; dendrites

1. Dendrites for Wordspotting

Dendrites are highly branched tree like structures that connect neuron's synapses to the soma. They were previously believed to act just like wires and have little or no computational value. However, studies show that dendrites are computational subunits that perform some inherent processing that contributes to overall neural computation [1–6]. It is thus interesting to explore computational models that can be built using dendrites as a unit. It has been shown that dendrites can perform computations similar to an HMM branch [3,7] which can be used for wordspotting. Wordspotting is the detection of small set of words in unconstrained speech [8]. The interlink between Neuroscience, CMOS transistors and HMMs is shown in Figure 1a.

Figure 1. (**a**) The Venn Diagram depicts the interlinks between the fields of neurobiology, HMM structures and CMOS transistors. We have demonstrated in the past how we can build reconfigurable dendrites using programmable analog techniques. We have also shown how such a dendritic network can be used to build an HMM classifier which is typically used for speech recognition systems; (**b**) Block Diagram for a Speech/Pattern Recognition system with respect to biology. In a typical speech recognition system, we have an auditory front-end processing block, a signal to symbol conversion block and a state decoding block for classification. We have implemented the state decoding block using dendritic branches, WTA and supporting circuitry for wordspotting. It is the classification stage before which symbols have been generated for a word.

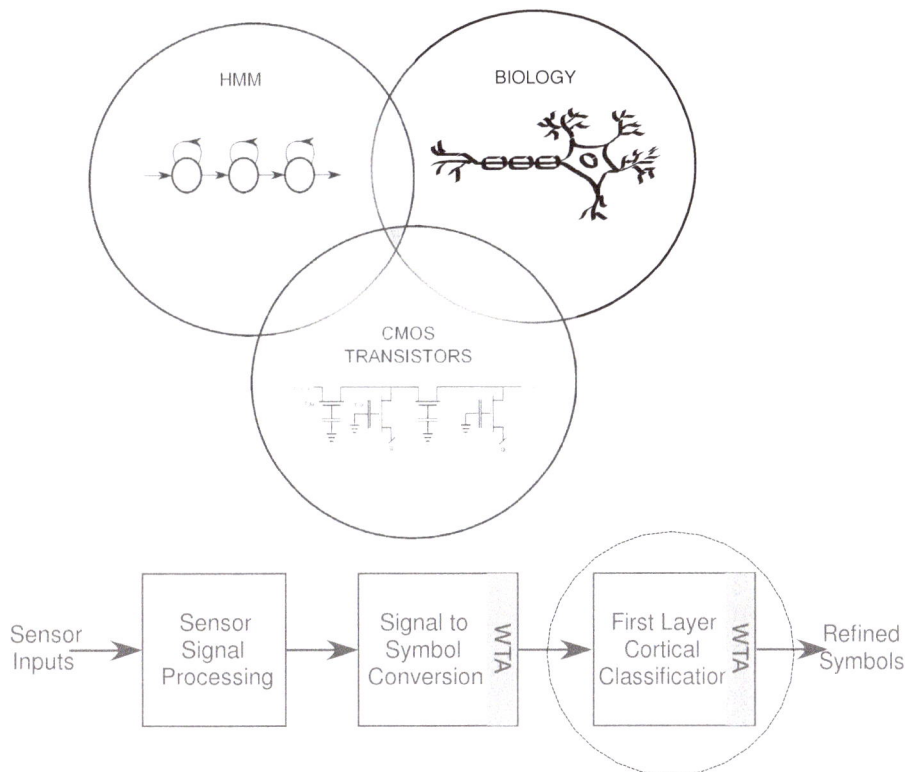

A typical Wordspotting system has at least three stages: Feature generation, Probability Generation (Signal to symbol conversion) and the State Decoding (classification) stage, which determines the word detected. Figure 1b shows the general block diagram for a classification system. In the specific example of speech recognizer, the sensor would be a microphone stage. The first stage has interface circuitry to acquire the signal from the microphone as well as initial signal processing. This processing may include signal conditioning and filtering, frequency decomposition as well as signal enhancement.

Figure 2 shows the FPAA as a prototyping device for audio signal processing applications. Our approach to audio processing includes a range of signal processing algorithms, that fit into the pathway between speech production (source) and perception (human ear). These algorithms are implemented by non-linear processing of sub-banded speech signals for applications such as noise suppression or hearing compensation, by proper choice of the non-linearity. In addition, the outputs of the non-linear processor can be taken at each sub-band, for speech detection instead of recombining to generate a perceptible signal for the human ear. Using this general framework, a variety of non-linear processing can result in applications in speech classifiers and hearing aid blocks. Here, we focus on the application of speech enhancement by noise-suppression, targeting word recognition in noisy environments. Detailed experimental results for a noise suppression application are discussed in [9], where the speech-enhanced sub-band signals are recombined together. For a speech recognizer, we use the enhanced sub-band signals directly to extract basic auditory features.

Figure 2. High level overview: The FPAA can be used for a variety of audio processing applications using the signal framework described. The first stage is a frequency decomposition stage followed by a non-linear processing block. The non-linear circuit can be used to implement the SNR estimator and a soft-threshold, which sets the gain in each sub-band. The gain control is implemented using a multiplier. Transient results from a MATLAB simulation of a 4 channel system is plotted. The noisy speech is gray, while the processed speech is in black.

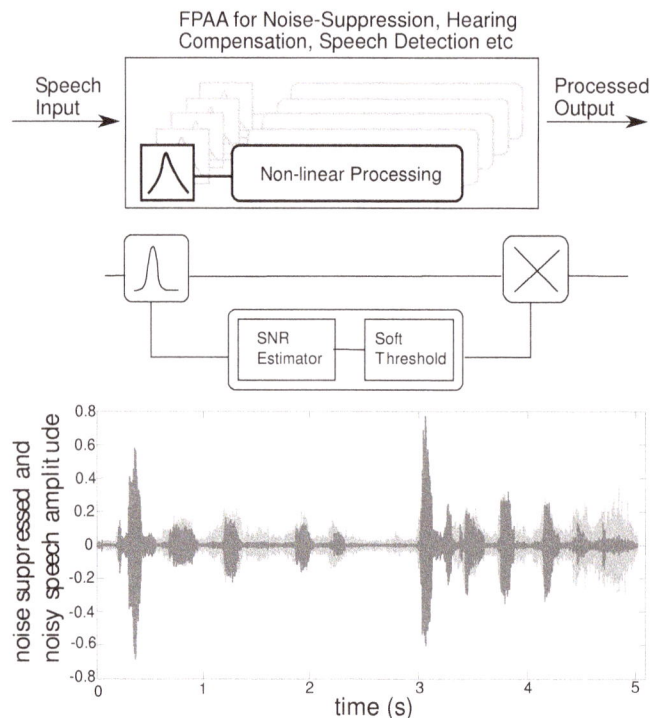

The second stage of the speech classifier consists of the probability generation stage that detects basic auditory features and supplies input probabilities to the state decoding stage. These enhanced sub-band signals undergo first-level information refinement in the probability generation stage, resulting in a sparse "symbol" or "event" representation. This stage maybe implemented as an Artificial Neural network (ANN), Gaussian Mixture model (GMM) or a Vector Matrix Multiplier (VMM) + WTA classifier. A typical 2-layer NN has synaptic inputs represented by the VMM and the sigmoid modeling the soma of a point-neuron. Alternatively, we can have synaptic computation followed by a competitive network modeled by the WTA.

We show in [10] that a single-stage VMM + WTA classifier can be used as a universal approximator, in contrast to an ANN implementation which requires two layers to implement a non-linear decision boundary. Figure 3 shows the comparison in circuit complexity of a two-layer ANN and a VMM + WTA classifier. A 1-layer NN requires the computation of a Vector-Matrix Multiply (VMM) + neuron. The addition of various weighted inputs is achieved through Kirchhoff's Current Law (KCL) at the soma node, adding all currents. The computation at the neuron is governed by the choice of complexity in the model. Usually, for moderate size of the network, the synaptic computation dominates the neuron computation. The sigmoidal threshold block for the soma nonlinearity in a NN can be implemented in voltage mode by converting the current output from the VMM into voltage and using a voltage-mode threshold block, or in current mode with an $arcsinh(.)$ block. Either of these implementations require more transistors per neuron compared to a WTA, which requires as few as 2 transistors per neuron.

Figure 3. Basic auditory feature extraction and probability generation stage: The speech input undergoes frequency decomposition or enhancement resulting in sub-band signals. The probability generation block can be implemented using an ANN, GMM or the VMM + WTA classifier. The circuit complexity is halved by using a VMM + WTA classifier.

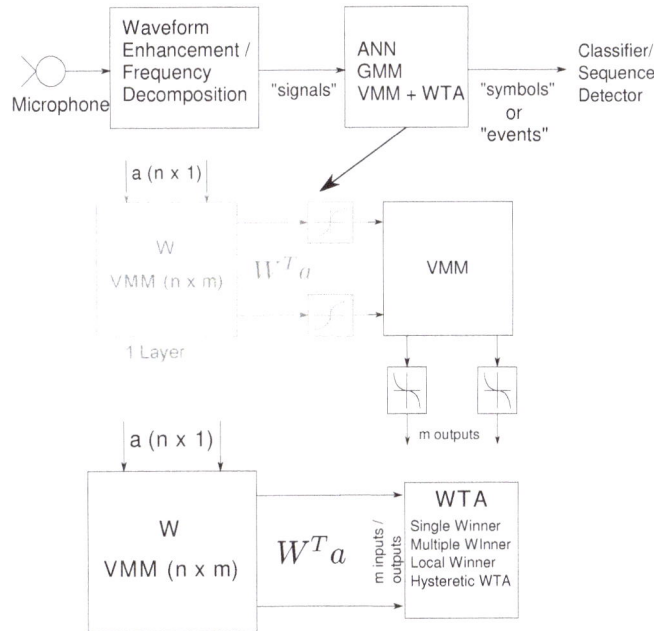

The VMM + WTA classifier topology has the advantage of being highly dense and low power. Each multiply is performed by one single transistor that stores the weight as well, and each WTA unit has only 2 transistors, providing very high circuit density. Custom analog VMMs have been shown to be $1000\times$ more power efficient than commercial digital implementations [11]. The non-volatile weights for the multiplier can be programmed allowing flexibility. The transistors performing multiplication are biased in deep sub-threshold regime of operation, resulting in high computing efficiency. We combine these advantages of VMMs with the reconfigurability offered by FPAA platforms to develop simple classifier structures.

In this paper, we demonstrate the state decoding stage of a simple YES/NO wordspotter. We have implemented an HMM classifier using bio-physically based CMOS dendrites for state decoding. For all experimental results in this paper, it is assumed that we have the outputs of the feature and probability generation stages.

We shall describe an HMM classifier model and its programmable IC implementation using CMOS dendrites. The first part of this paper describes the similarity between a single dendritic branch and HMM branch, in addition to exemplifying its usage to compute a metric for classification. An HMM classifier is modeled comprising of these dendritic branches, a Winner-Take-All (WTA) circuit and other supporting circuitry. Subsequently, the computational efficiency of this implementation in comparison to biological and digital systems is discussed. Intriguingly, this research substantiates the propensity of computational power that biological dendrites encompass, allowing speculation of several interesting possibilities and impacts on neuroscience. It is in some ways a virtual visit into the dendritic tree as was suggested by Segev *et al.* [12]. This paper further explores the interlinks between neurobiology, Hidden Markov Models and CMOS transistors based on which we can postulate that a large group of cortical cells function in a similar fashion as an HMM network [4,7]. Section II describes the similarities between a dendrite branch and an HMM branch. We discuss the similarities between a simulated HMM branch and experimental results using a CMOS dendrite branch. In Section III, we discuss the single CMOS dendrite in detail. We will present experimental results for the line for different parameters. We also discuss the simulation model that we have developed and the similar results seen. In section IV, we discuss the Analog HMM classifier implementation. We discuss the experimental results for a YES/NO wordspotter for different sequences. In section V, we discuss the tools that made the implementation of this classifier structure possible. In section VI, we will discuss the computational efficiency of the system as compared to digital and biological systems. In the final section we will summarize the results and discuss the future possibilities.

2. Dendritic Computation and the HMM Branch

For a typical HMM used for speech recognition, the update rule is given by:

$$\phi_i[n] = b_i[n]((1 - a_i)\phi_i[n-1] + a_{i-1}\phi_{i-1}[n-1]) \tag{1}$$

The probability distribution $b_i[n]$, represents the estimate of a symbol (short segment of speech/phoneme) produced by a state i in frame n. $\phi_i[n]$ represents the likelihood that a particular state, was the end-state in a path of states that models the input signals [13] as shown in Equation (1). a_i is the transition probability from one state to another. In a typical speech recognition model, the states would be phonemes/words and the output would represent the audio signal produced by the subject. The features of the audio signal tend to vary for different subjects. The goal of this classifier model is to correctly classify a sequence of symbols with some tolerance. For an HMM state machine for speech recognition using CMOS dendrites, the inputs $b_i[n]$ can be modeled as Gaussian inputs as shown in Figure 4a, which is typical for $b_i[n]$ for speech signals with an exponential rise-time and fall-time. In Figure 4b, the likelihood outputs for each state shows a very a sharp decay and has a very high dynamic range.

Figure 4. Simulation results for an HMM state machine based on a Mathematical HMM model built using MATLAB (**a**) Input probability distribution of different symbols varying with time; (**b**) Likelihood outputs of all the states on a logarithmic scale; (**c**) Normalized likelihood outputs of all the states. The outputs were normalized by multiplying them with an exponential function of the form $exp(n/\tau)$.

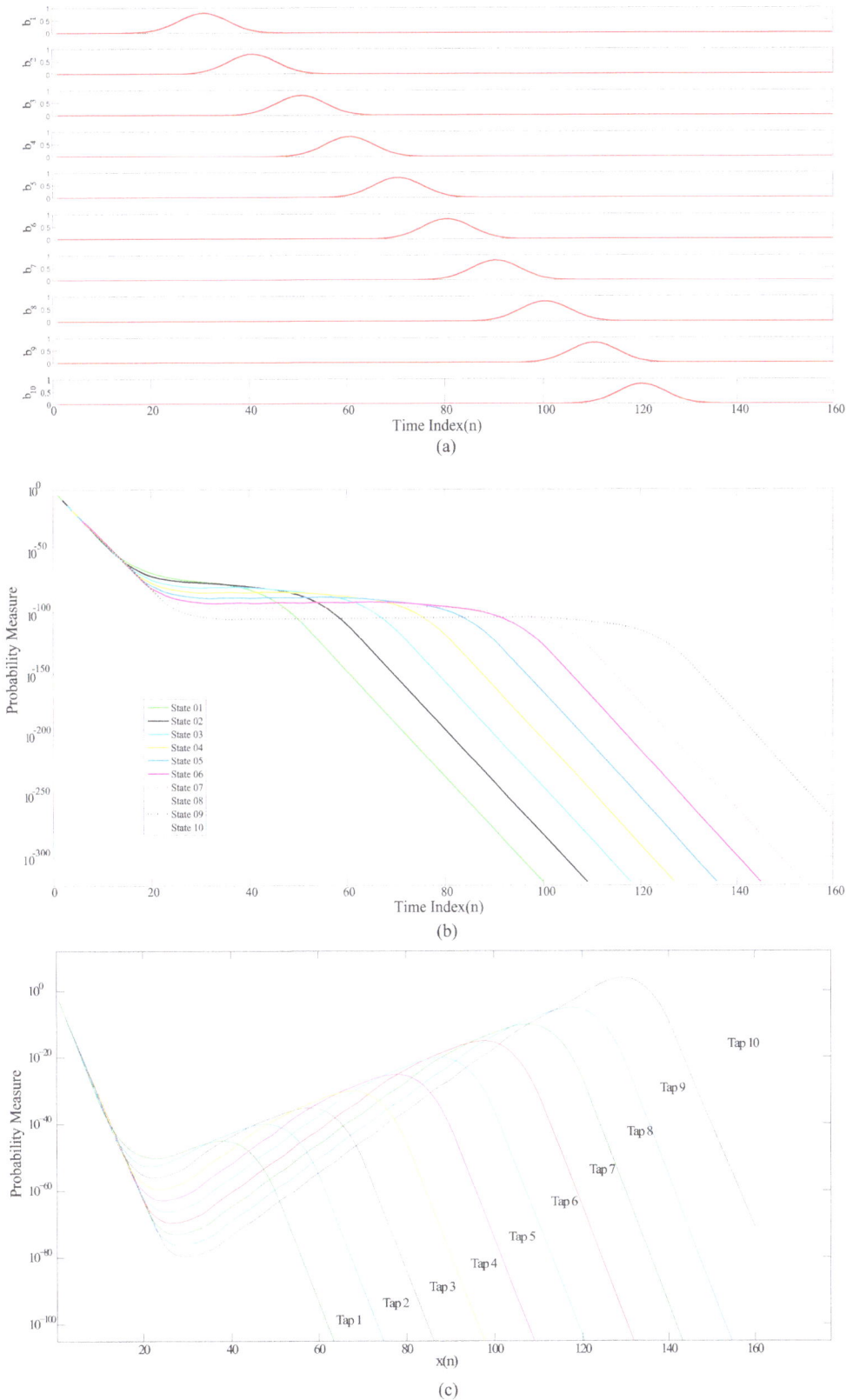

Figure 5. CMOS implementation for a dendritic branch and experimental results. (**a**) Dendrite with increasing diameter as typically seen in pyramidal cells. We refer this increasing diameter as "taper"; (**b**) Co-relation between basic left-to-right HMM branch and a CMOS dendrite branch with "taper"; (**c**) Resulting IC implementation using programmable analog floating-gate pFETs. For the CMOS dendrite the "taper" is modeled by increasing the axial conductance from left-to-right; (**d**) Experimental results showing the outputs from each tap of the CMOS dendrite. These outputs are equivalent to likelihood outputs from the HMM states. The output doesn't decay completely but attains a new dc level. Note that we did not do normalization explicitly for the outputs of the dendrite as the decay is not as sharp as seen in HMMs. All taps are set initially to have the same membrane voltage V_{mem}.

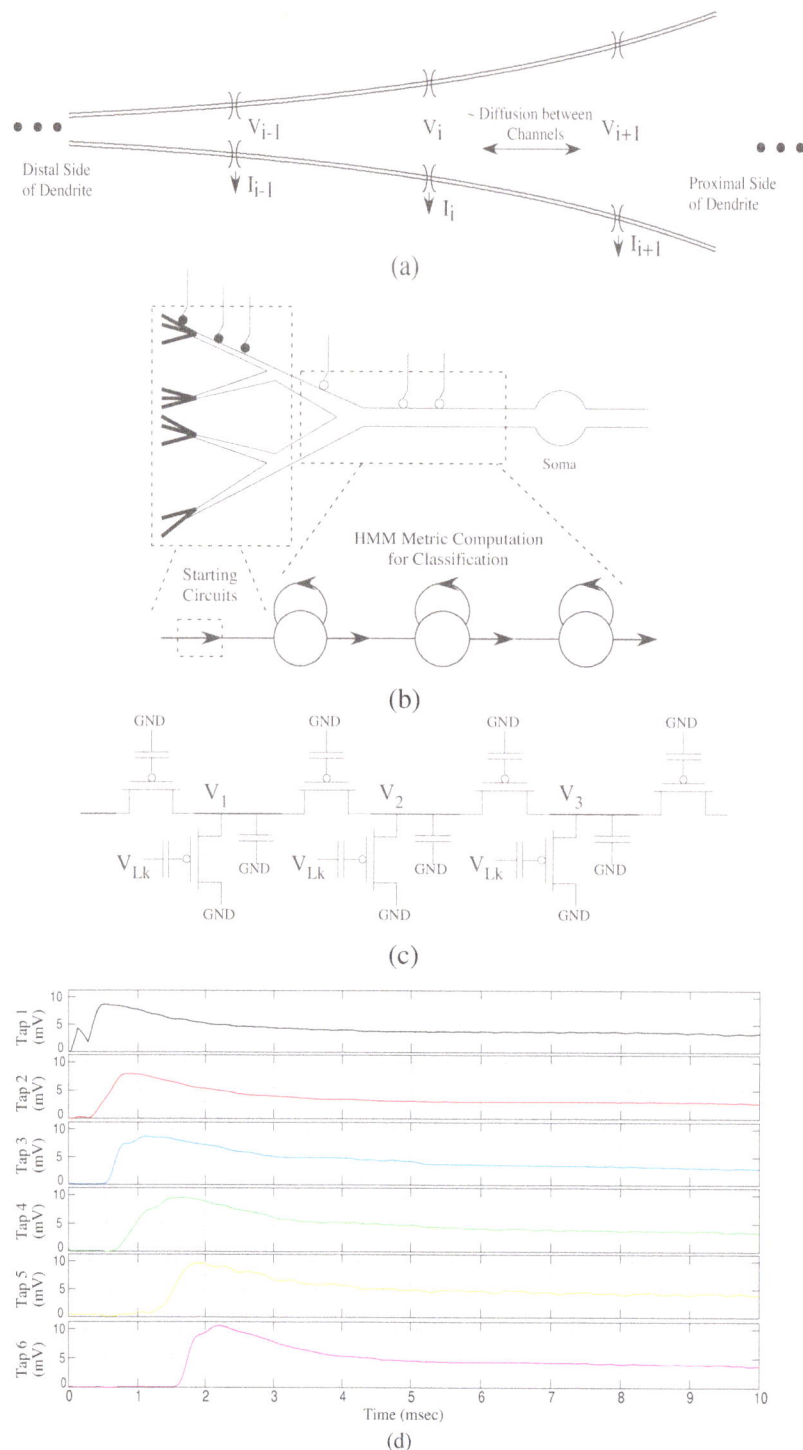

(a)

(b)

(c)

(d)

To limit this range, we normalize this output with an exponential function. It can be observed that the normalized likelihood is similar to an EPSP signal with an asymmetric rise and fall time. For a single n-stage dendritic line with "taper", if we applied sequential EPSP inputs at subsequent nodes, the output observed at the end of the line is as shown in Figure 5d. A "taper" signifies the changing diameter through the length of a dendrite. It represents the normalized likelihood outputs of an HMM classifier. The Gaussian inputs for the HMM model can be modeled using synaptic currents for a dendrite which is also typical for biological systems. $b_i(t)$ is thus represented as the synaptic current into each node. The output voltage of each tap of the dendrite represents the likelihood $\phi_i(t)$ of an HMM state. This can be linearly-encoded or log-encoded depending on the region of operation. For the dendritic system, no normalization is done as the decay is not as sharp as seen in the HMM branch for a wide dynamic range.

For a continuous-time version of Equation (1), the update rule is given by,

$$\phi_i(t) = b_i(t)((1 - a_i)\phi_i(t - \tau) + a_{i-1}\phi_{i-1}(t - \tau)) \tag{2}$$

where, $b_i(t)$ is the input probability of symbol in state i; and $\phi_i(t)$ is the likelihood of a state i at time t; τ is the time index between two consecutive time indexes and a_i is the transition probability between adjacent states. Even though the state sequence is implied, one cannot assume a definitive observation of transition between the states. This is the reason why it is called *Hidden* Markov Model although the state sequence has a Markovian structure [14]. Continuous-time HMMs can be represented as a continuous-time wave-propagating PDE as given in Equation (3) [15].

$$\underbrace{\tau \frac{\partial \varphi(x,t)}{\partial t}}_{\substack{\text{state} \\ \text{element}}} + \underbrace{\left(\frac{1}{b(x,t)} - 1\right)\varphi(x,t)}_{\substack{\text{decay} \\ \text{term}}} + \underbrace{a(x)\Delta\frac{\partial \varphi(x,t)}{\partial x}}_{\substack{\text{wave} \\ \text{propagation}}} = 0 \tag{3}$$

where, Δ is the distance between two state nodes. This can be compared to analog diffuser circuits. Also, an HMM branch and a dendrite branch have similar looking topologies and similar wave-propagating properties. The HMM state machine used, as shown in Figure 4a, is a left-to-right model. Studies have shown that a biological dendrite also does not have a constant diameter [16]. Its diameter at the distal end is smaller as compared to the proximal end as shown in Figure 5a,b [17]. Thus, for a similar CMOS dendritic line that is uni-directional, we would expect the axial conductances of the line to increase from left-to-right as shown in Figure 5c. This is the case of a dendrite with "taper". Such a topology ensures that the current flow is uni-directional. This also favors coincidence detection in the dendrite. We can compare the continuous-time HMM to an RC delay line with "taper". For this let us analyze the behavior of an RC delay line with and without taper.

2.1. RC Delay Line without Taper

The classical RC delay line is reviewed in Mead's text [18]. Figure 6 shows the topology. Kirchhoff's Current Law (KCL) can be used to derive a differential equation for this circuit, given by Equation (4), where G is conductance.

Figure 6. RC delay line representing a dendrite. The Rs represent the axial resistances, the Gs represent the leakage conductances and C is the membrane capacitance.

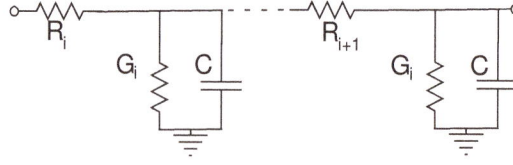

$$I_i\left(t\right) = C_i\frac{dV_i(t)}{dt} + V_i\left(t\right)G_i + \frac{[V_i(t)-V_{i+1}(t)]}{R_{i-1}}$$
$$+ \frac{[V_i(t)-V_{i-1}(t)]}{R_i} \tag{4}$$

Assuming the horizontal resistances are equal as given in Equation (5) allows one to simplify Equation (4) to Equation (6):

$$R_i = R_{i-1} = R_x \tag{5}$$

$$I_i\left(t\right) = C_i\frac{dV_i(t)}{dt} + V_i\left(t\right)G_i$$
$$+ \frac{1}{R_x}\left[2V_i\left(t\right) - V_{i+1}\left(t\right) - V_{i-1}\left(t\right)\right] \tag{6}$$

Assuming there are many nodes allows one to perform the following change of notation from discrete nodes to continuous nodes:

$$V_i(t) = V(x,t) \tag{7}$$
$$V_{i+1}\left(t\right) = V\left(x + \Delta_x, t\right) \tag{8}$$
$$V_{i-1}\left(t\right) = V\left(x - \Delta_x, t\right) \tag{9}$$

Assuming that Δ_x represents a "position delta" one may use the Taylor series to describe the continuous nodes in terms of Δ_x, Equations (10) and (11).

$$V\left(x + \Delta_x, t\right) = V\left(x,t\right) + \Delta_x\frac{dV\left(x,t\right)}{dx} + \frac{1}{2}\left(\Delta_x\right)^2\frac{d^2V\left(x,t\right)}{dx^2} + \cdots \tag{10}$$

$$V\left(x - \Delta_x, t\right) = V\left(x,t\right) - \Delta_x\frac{dV\left(x,t\right)}{dx} + \frac{1}{2}\left(\Delta_x\right)^2\frac{d^2V\left(x,t\right)}{dx^2} + \cdots \tag{11}$$

Substituting Equations (10) and (11) into Equation (6) and simplifying, yields Equation (12), the generalized PDE describing the RC delay line diffusor.

$$I_i\left(t\right)R_x = R_xC_i\frac{dV_i\left(t\right)}{dt} + R_xG_iV_i\left(t\right) - \left(\Delta_x\right)^2\frac{d^2V\left(x,t\right)}{dx^2} \tag{12}$$

If one assumes no input current at the top of each node $I_i = 0$, then one can put the diffusor circuit into a form similar to the continuous time HMM equation as given in Equation (13).

$$\underbrace{R_xC_i\frac{dV\left(x,t\right)}{dt}}_{\substack{state \\ element}} + \underbrace{R_xG_iV\left(x,t\right)}_{\substack{decay \\ term}} - \underbrace{\left(\Delta_x\right)^2\frac{d^2V\left(x,t\right)}{dx^2}}_{\substack{diffusion \\ term}} = 0 \tag{13}$$

The impulse response of such a system is a Gaussian decaying function over time. In this case, diffusion is the dominant behavior of the system.

2.2. RC Delay Line with Taper

Assuming that HMM will always propagate to the next state and there is no probability that it will remain in its current state leads to the assumption as given in Equation (14) which can be substituted in Equation (3):

$$a\left(x\right) = 1 \tag{14}$$

For a dendrite circuit with taper, axial conductances are NOT equal and increase towards the right. Using this assumption, Equation (4) simplifies to Equation (15):

$$I_i\left(t\right) = C_i \frac{dV_i\left(t\right)}{dt} + V_i\left(t\right)\left[G_i + \frac{1}{R_i}\right] - \frac{V_{i-1}\left(t\right)}{R_i} \tag{15}$$

Substituting the Taylor series expansions of Equations (10) and (11) into the above we get:

$$
I_i\left(t\right) = C_i \frac{dV(x,t)}{dt} + V\left(x,t\right)\left[G_i + \frac{1}{R_i}\right]
\\
- \frac{1}{R_i}\begin{bmatrix} V\left(x,t\right) \\ -\Delta_x \frac{dV(x,t)}{dx} \\ +\frac{1}{2}(\Delta_x)^2 \frac{d^2V(x,t)}{dx^2} \end{bmatrix} \tag{16}
$$

Assuming that

$$\Delta x \ll 1 \tag{17}$$

we can neglect higher order terms of the Taylor series.

$$(\Delta_x)^2 \approx 0 \tag{18}$$

We can see in Equation (16) that there is still some diffusion that can be seen in the line. It is however negligible as the wave propagation term is more dominant. Re-arranging terms and assuming no input current we get:

$$0 = \underbrace{R_i C_i \frac{dV\left(x,t\right)}{dt}}_{\substack{state \\ element}} + \underbrace{V\left(x,t\right)\left[G_i R_i - 1\right]}_{\substack{decay \\ term}} + \underbrace{\Delta_x \frac{dV\left(x,t\right)}{dx}}_{\substack{wave \\ propagation}} \tag{19}$$

Table 1 closely examines the similarities between a RC delay line and an HMM PDE.

Table 1. Comparing HMM PDE and RC Delay Line Terms w/Assumptions.

Element description	HMM PDE	RC delay line
Recursion variable	$\varphi\left(x,t\right)$	$V\left(x,t\right)$
State element coefficient	τ	$R_i C_i$
Decay term coefficient	$\frac{1}{b(x,t)} - 1$	$G_i R_i - 1$
Wave propagation/diffusion term	$K \frac{\partial \varphi(x,t)}{\partial x}$	$K \frac{dV(x,t)}{dx}$

3. Single Line CMOS Dendrite

Since dendrites have computational significance, it is interesting to explore computational models that can be built using dendrites or a network of dendrites. One such application is classification in speech recognition. We have already discussed the similarities between an HMM branch and a dendritic branch. To test this hypotheses, we implemented a single dendritic branch with spatially temporal synaptic inputs. We compared a single CMOS dendritic branch implemented on a reconfigurable analog platform and a MATLAB Simulink simulation model based on the device physics of CMOS transistors. Figure 7 shows a complete overview of how CMOS dendrites are modeled and also the experimental results for a 6-compartment CMOS dendrite. The inputs to the dendrite are synaptic currents. In biological systems, synaptic inputs can be excitatory and inhibitory in nature. However, in this paper we assume that we have excitatory synapses as a majority of contacts on a pyramidal cell are excitatory in nature. As discussed before the dendrite does not have a constant diameter. This implies that for a CMOS dendrite, the conductance of the dendrite increases towards the soma *i.e.*, from left to right [17]. The inputs will also decrease in amplitude as conductance increases. This ensures that an input closer to the soma does not have a larger effect than inputs farther away. This indicates decreasing synaptic strengths of inputs down the dendritic line. This has been observed previously in biological dendrites [16]. Thus, we also varied the synaptic strengths of inputs in our experiments. We implemented the single dendritic line both as a CMOS circuit model and a MATLAB Simulink simulation model. We found that the comparison of our experimental and simulation results were fairly close. This is demonstrated in Figure 8.

Figure 7. System overview for a dendrite branch. (**a**) Detailed diagram for a single dendritic line which is equivalent to an HMM branch; (**b**) The representation of input voltage on the source of the transistor representing the input synapses; (**c**) The asymmetric triangular input voltages V_{syn} on the source of the transistor representing the input synapses. I_{syn}, the input synapse currents into each of the different nodes is proportional to V_{syn}; (**d**) V_{ota}, the output of FG-OTA which has a gain of approximately 20; (**e**) V_{out}, the output voltage at each node.

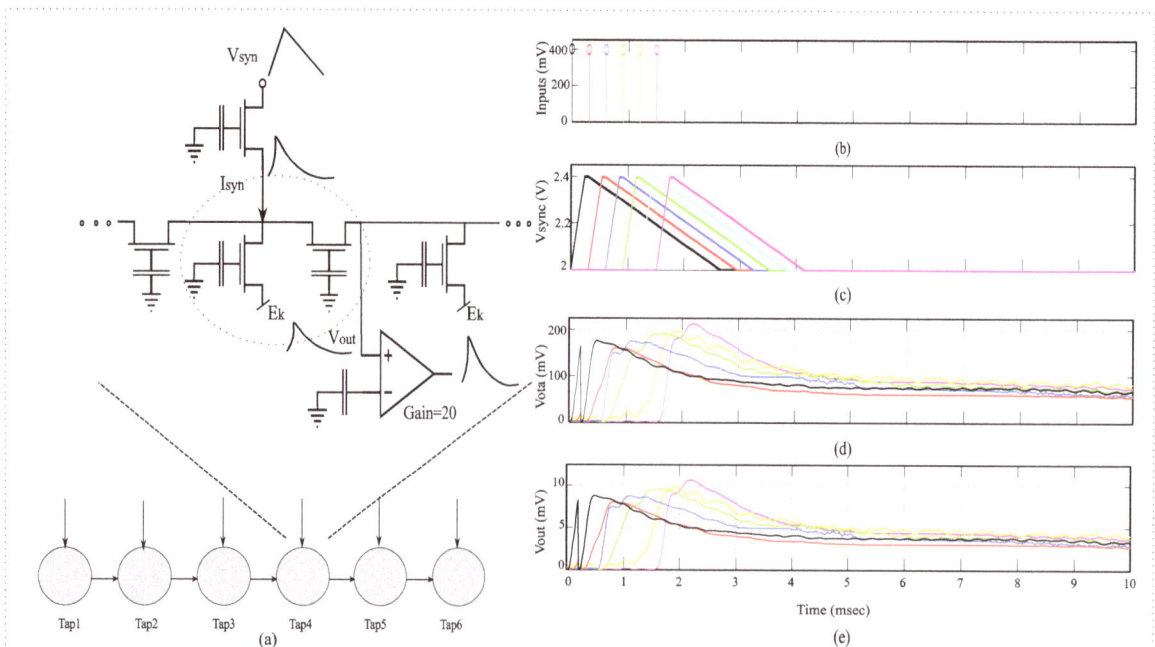

Figure 8. Simulation Data *vs.* experimental data comparison. The dotted lines depict the simulation data and the solid lines are the experimental data. The parameters for simulation data are $V_{Leak} = 0.5V$, $V_{axial} = 0.5V$, $\kappa = 0.84$, $I_0 = 0.1fA$, $C = 1.3pF$, $E_k = 1V$, $V_{dd} = 2.4V$.

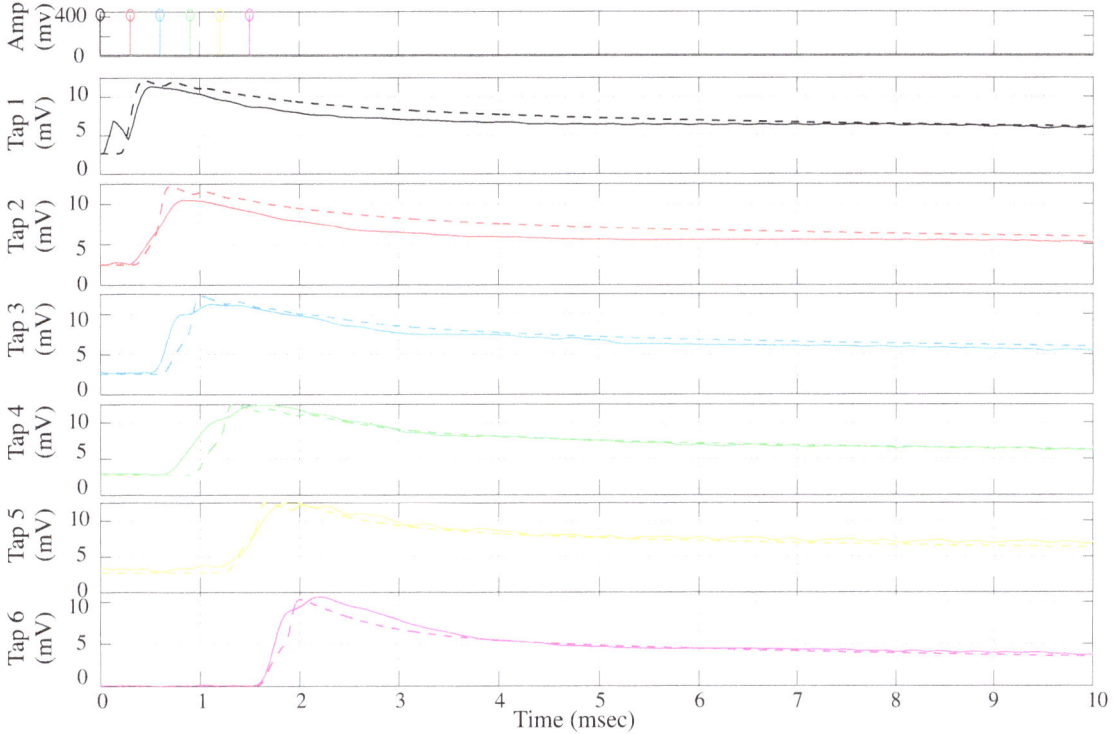

3.1. Inputs to the PFET Source

The input probabilities $b_i(t)$ are represented as log-compressed voltage signal at the dendrite node. To generate EPSP input currents into each of the dendritic nodes, we input an asymmetric triangular wave voltage at the source of the pFET FG-FETs. This generates typical EPSP signals, which have a faster rising time and a slower fall time. By varying the magnitude of the triangular waves we were able to control the input current into each of the nodes of the dendrite. This can be seen in Figure 7c. The current of a transistor is exponentially proportional to its source voltage V_S.

$$I_{syn} = I_0 e^{\kappa(V_S - V_G)/U_T} \left(e^{-(V_S - V_D)/U_T} - 1 \right) \tag{20}$$

where, $V_S = V_{dd}$. This enables us to generate EPSP-like inputs for the CMOS dendrite. All input representations shown thus are voltage inputs on the source of the transistor, that acts as synapse at every node of the dendrite.

3.2. Single Line Dendrite Results

We implemented a single 6-compartment dendrite. Each compartment consisted of 3 FG pFETS for the axial conductance, the leakage conductance and the synaptic input respectively. We present experimental results for the same. To test the behavior of dendrites for a typical speech model, we varied three parameters namely: "taper", delay between inputs and the EPSP strengths of the synaptic inputs. In terms of "taper", two approaches were tested. One without "taper" and the second with increasing "taper". Results are shown in Figure 9a. We observed that by using "taper" we could ensure that the input current would transmit more in one direction of the dendritic cable. To achieve this we increased the axial conductance of the cable down the line, such that maximum current tends to flow to the end of the cable. At every node of the dendrites we input EPSP currents in a sequence. This is similar to a speech processing model, where all the phonemes/words are in a sequence and based on the sequence we classify the word/phoneme. We then varied the delay between the input EPSP signals as seen in Figure 9b. It was observed that as the delay between the inputs increases, the amplitude of the output decreases. This implies that as outputs are spaced farther apart, there is less coincidence detection. The third parameter varied was the strength of the EPSP inputs, with the difference in EPSP strengths of the first node and the last node increasing for subsequent plots as seen in Figure 9c. The EPSP strengths near the distal end are larger than the EPSP strengths near the proximal end. Evidence for the same has been shown in biology [16]. It was observed that as the difference in amplitude was increased, the amplitude of the output reduced. The study of the variation of these parameters showed the robustness that such a system would demonstrate in terms of speech signals. The difference in delay, models the different time delays between voice signals when a word is spoken by different subjects. The difference in EPSP strengths ensures that the impact of all the phonemes on the output is similar for detection of a word and not dominated by just the last stage.

In Figure 10, we have studied the trends that one would observe collectively for different parameters. The output metric here is the difference of amplitude of last node when all inputs are present and when only the last input is present. We observed that as we increased the timing difference between various inputs, the final metric of the line decreased as seen in Figure 10b. We simulated the dendritic branch to observe the effects a wide range of time delays between inputs as shown in Figure 10c. We observed that the output metric decreased as we increased the delay between the inputs for a line. And for the cases where we reversed the sequence, the amplitude was very close to zero. This clearly demonstrates that if the sequence of the inputs is not in succession, there will be no word detection. Also, the output metric decreases as the delay between the inputs increases.

Figure 9. Experimental results for a single branch 6-tap dendrite for different parameters. The three main parameters that govern the output of a dendrite are, namely the taper of the line, the spatial-temporal characteristics of the synaptic inputs and the strength of the synaptic inputs. All results are from the last tap of the dendrite. (**a**) Metric changed is the taper of the dendrite. For subsequent figures, the taper is increased from no taper to a larger taper. The diameter of the dendrite increases down the line which is achieved by increasing the conductances of the axial transistors from left to right; (**b**) Metric changed is the delay between EPSP inputs into each of the taps of the dendrite. In the first case we have zero time unit delay, 10 time units delay $(2\,ms)$ for second and 20 time units delay $(4\,ms)$ for the third diagram in the sequence. One time unit = $0.2\,ms$; (**c**) Metric changes is the difference between the EPSP strengths of the input signals. In the first case, the difference is 10 mV, 50 mV for the second and 100 mV for the third case. As can be seen in the graph we can see decreasing amplitude as the difference in EPSP strengths increases

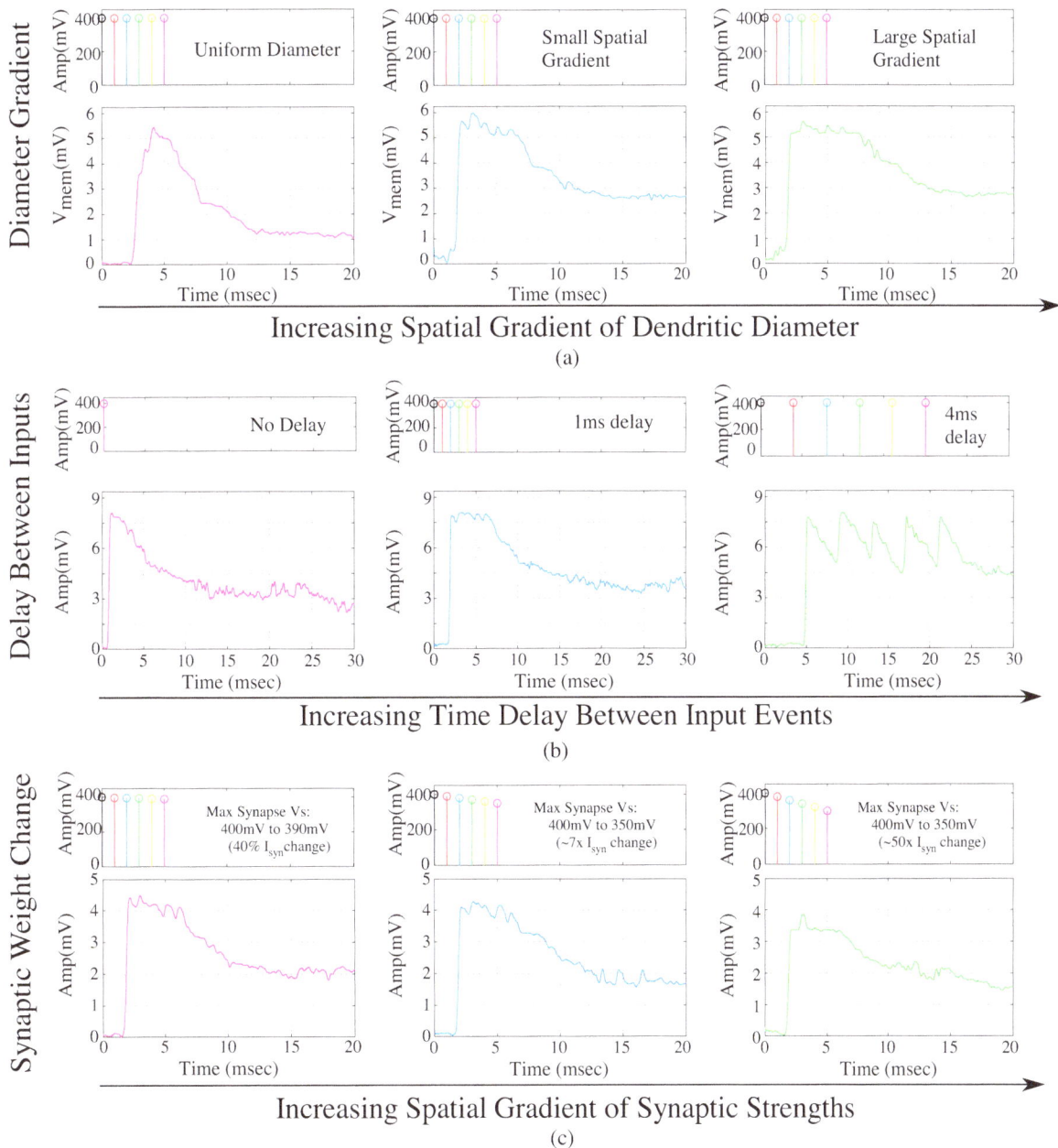

Figure 10. Experimental results, simulation results and trends observed for a single line dendrite. We varied the input sequence with respect to the time difference between signals. The output metric in this case is the difference between the output of the dendrite when all signals were present and output of the dendrite when only the last input was present. (**a**) Diagram depicting the decreasing EPSP inputs into a single CMOS dendrite line; (**b**) Experimental data showing change in peak to peak amplitude for a dendrite as the EPSP inputs into each of the nodes decrease down the line; (**c**) Change in amplitude of the output with respect to increasing difference in the EPSP amplitudes as we progress from left to right down the line. t_{diff} implies the time delay between inputs. As we increase the time delay the output metric reduces. Negative t_{diff} implies a reversed sequence of inputs, where the output metric is zero; (**d**) Change in amplitude of the output with respect to increasing difference in the taper of the dendrite. In this experiment, the diameter of the dendrite was increased as we progress from left to right down the line. t_{diff} implies the time delay between inputs. As we increase the time delay the output metric reduces. Negative t_{diff} implies a reversed sequence of inputs, where the output metric is zero. The study of these parameters showed the robustness that such a system would demonstrate in terms of speech signals. The difference in delay, models the different time delays between voice signals when a word is spoken by different people. The difference in EPSP strengths ensures that the impact of all the phonemes on the output is similar for detection of a word and not dominated by just the last stage.

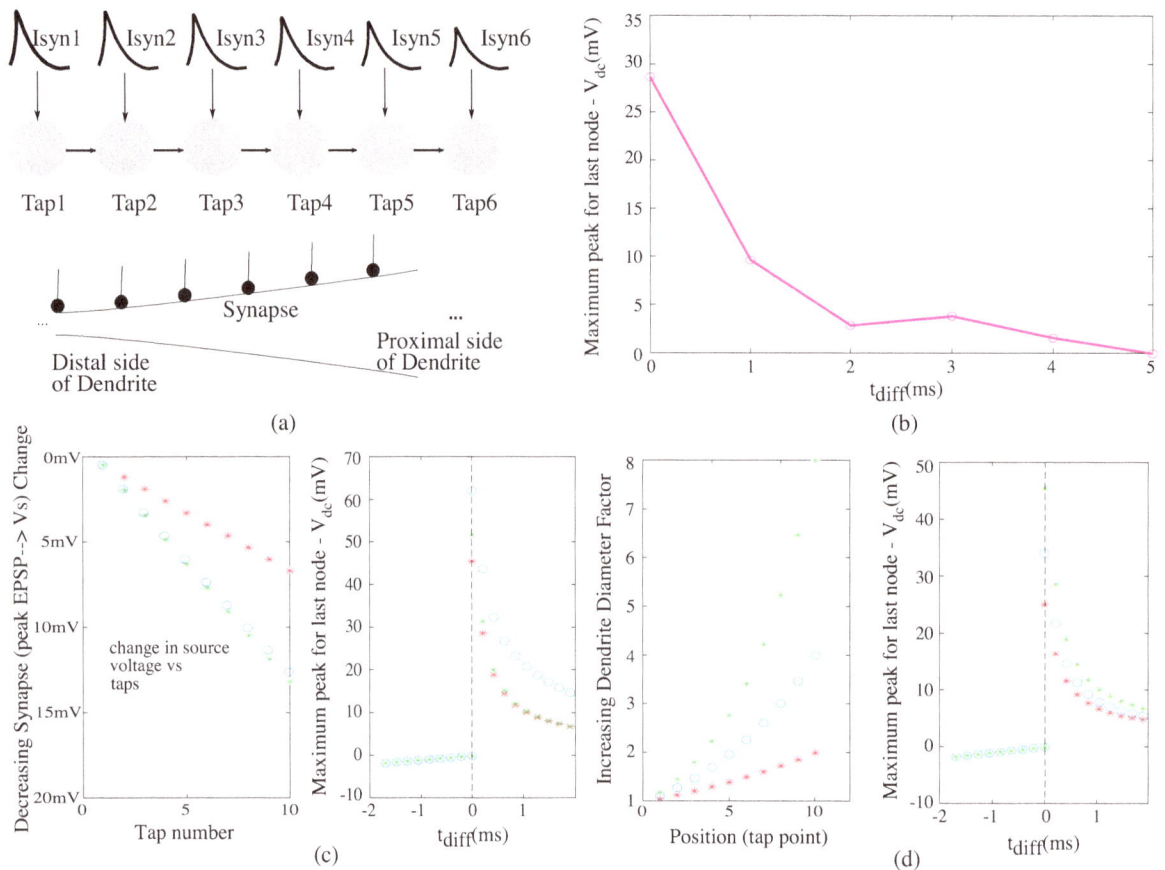

4. Analog Classifier for Word-Spotting

We will now discuss the complete classifier structure. We have built a simple YES/NO HMM classifier using dendrite branches, a Winner-Take-All (WTA) circuit and supporting circuitry. We will simplify the modeling of a group of neuron somas and the inhibitory inter-neurons as a basic WTA block, with one winning element. We can consider the winning WTA output, when it transitions to a winning response as an equivalent of an output event (or action potential). To build this network, we made a model of a dendrite, initially a single piece of cable with branch points, where the conductance of the line gets larger towards the soma end, and the inputs are excitatory synaptic inputs. For classification, we focus on the ability for dendritic trees to be able to compute useful metrics of confidence of a particular symbol occurring at a particular time. This confidence metric will not only be a metric of the strength of the inputs, but also will capture the coincidence of the timing of the inputs. We would expect to get a higher metric if the $1st$, $2nd$, and $3rd$, inputs arrived in sequence, whereas we would expect a lower metric for the $3rd$, $2nd$, and $1st$ inputs arrived in sequence. This type of metric building is typical of HMM type networks. Simple example being if the word "Y" "E" "S" were detected in a sequence as opposed to "S" "E" "Y". This is demonstrated by the simulation results as shown in Figure 10, where when the input sequence is reversed the output metric is zero. The output metric is defined as the difference in output of last node when all inputs are present and when only the last input is present.

Figure 11. (a) The classifier structure with the normalization factor multiplied, $f(t) = e^{t/\tau}$; (b) The classifier structure after normalization. This figure demonstrates that the normalization is inherent in the system; (c) Detailed structure of the HMM classifier using reconfigurable floating-gate devices. There are three main structures here : The dendrite branches, the Winner-Take-All circuit and the supporting circuitry. The dendrite branch consists of a 5-stage dendrite for both the branches representing the words YES and NO; and a single stage dendrite to set the the threshold current. The dendrites have synaptic inputs at each node, which represent the phonemes of the word to be detected. When the output of a dendrite exceeds the threshold limit *i.e.*, if a YES/NO is detected, the threshold loses. The supporting circuitry consists of a Vector-Matrix Multiplier (VMM) building block which acts as a reset after a word is spotted [11].

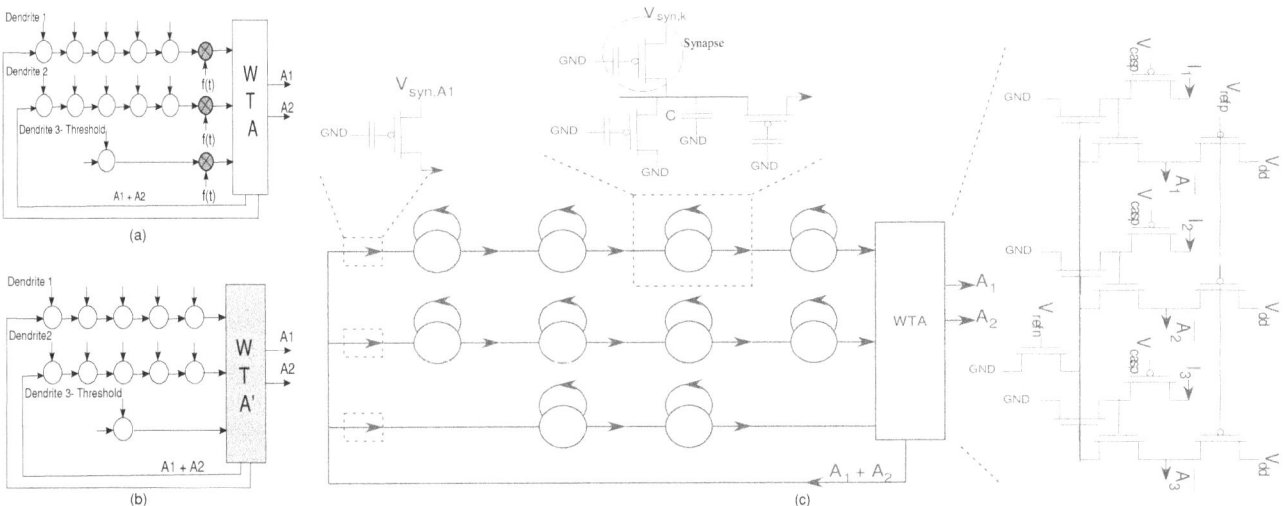

The network we built has two desired winning symbols, "YES" and "NO". Each symbol is represented by one or more states that indicate if a valid metric has been classified. Only the winning states would be seen as useful outputs. The useful outputs feed back to the input dendrite lines, and in effect reset the metric computations. This is implemented using a Vector-Matrix-Multiplier block [11]. The system block diagram is as shown in Figure 11. Each of the dendritic lines for the desired winning symbols has 5 states (dendritic compartments), where the inputs to the dendritic line represent typical excitatory synaptic inputs.

4.1. Synaptic Inputs Model Symbol Probability

In speech/pattern recognition, signal statistics/features are the inputs to the HMM state decoder. It generates the probability of the occurrence of any of the speech symbols. These signals when grouped, generate a larger set of symbols like phonemes or words [13]. We assume we have these input probabilities to begin with, as inputs to the classifier structure. We have taken inspiration from Lazzaro's Analog wordspotter for classification. However, we use a different normalization technique to eliminate the decay as shown in Figure 4c. We can draw comparisons for such a system to a biological dendrite with synaptic inputs. We have modeled the input signals as excitatory synaptic currents. The synaptic current is given by :

$$I_{syn} \propto te^{-t/t_{peak}} \tag{21}$$

For a continuous cable,

$$\tau \frac{dV(x,t)}{dt} + V(x,t) = \lambda^2(x)\frac{d^2V(x,t)}{dx^2} + R(x)I_{input} \tag{22}$$

Considering that $exp(t/\tau)$ is the normalizing factor we have,

$$V(x,t) = V_1(x,t)e^{t/\tau} \tag{23}$$

where,

$$\tau \frac{dV_1}{dt} + V_1(x,t) = \lambda^2(x)\frac{d^2V_1}{dx^2} + R(x)I_{input}e^{-t/\tau} \tag{24}$$

$V_1(x,t)$ is the system output before normalization. From Equations (21) and (24), we see that the input is similar to a synaptic current. Thus the inputs for the classifier using dendrites can be modeled as synaptic currents. This is represented in Figure 11a and Figure 11b. The derivation has two implications. First, we can use EPSP inputs to represent the input probabilities for phonemes. Second the system inherently normalizes the outputs. In Figure 12, the input to dendrite-1 signifies the phonemes of the word "YES". The inputs used were EPSP inputs that are similar to probability inputs $b_i(t)$ that in a typical HMM classification structure would be generated by a probability estimation block. There is no input into dendrite-2 which signifies that phonemes of "NO" were not detected. The threshold dendrite, dendrite-3 sets the threshold level. The WTA circuit determines the winner amongst the three dendritic lines. It is observed that when "YES" is detected, dendrite-1 wins. This happens when coincidence detection is observed at the output of dendrite-1. The winning line signifies the word that is classified. It is only when all the inputs are in sequence and cross the given threshold that the dendrite line wins. In Figure 12 we demonstrate the classification of the word "YES". The feedback from the WTA acts as a reset function for the dendrites, as after a word has been classified the threshold dendrite wins again. In Figure 13, the classification of words "YES" and "NO" in a sequence is demonstrated. In Figure 14 we show the effect of timing and variation of EPSP strengths for input sequences.

Figure 12. Experimental results for the YES/NO classifier system. The results shown are for the case when a YES is detected by the system (**a**) Synaptic inputs at the nodes of the first dendrite and the line output for the first dendrite. Here we assume we have the input probability estimate for the phonemes (symbols) for the word YES; (**b**) Corresponding WTA output for first dendrite. A low value signifies that it is winning; (**c**) The synaptic input and output for the second dendrite; (**d**) Corresponding WTA output for the second dendrite; (**e**) The line output for the third dendrite; (**f**) Corresponding WTA output of the third dendrite. The third dendrite acts as a threshold parameter. The amplitude of the word detected on a particular line needs to be higher than the threshold to win.

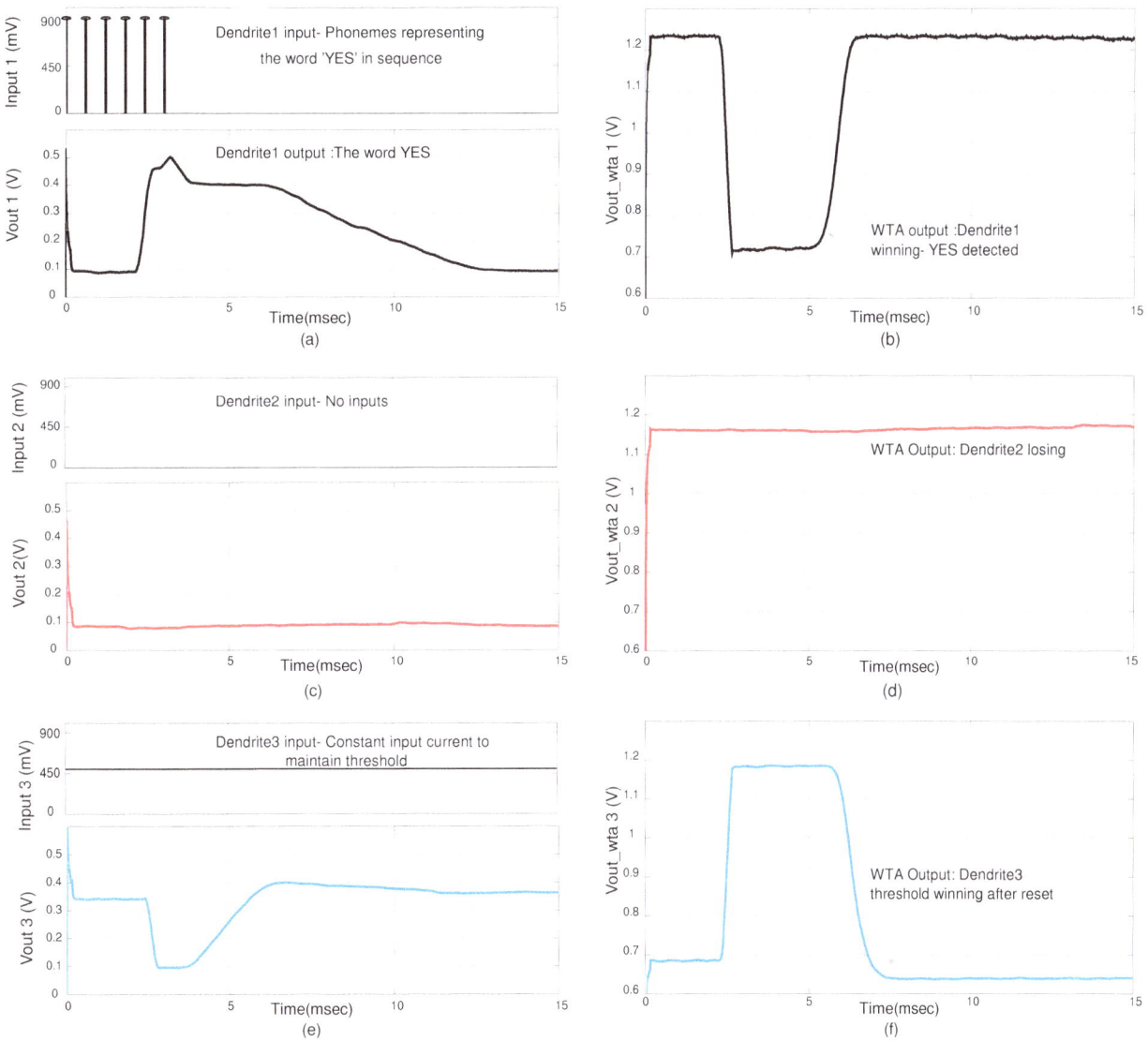

Figure 13. Experimental results for the classifier system when a sequence of words is detected. (**a**) First dendrite wins when the word YES is detected and the second dendrite wins when the word NO is detected. The WTA inputs and outputs are shown; (**b**) Second dendrite wins when the word NO is detected and first dendrite wins when YES in detected.

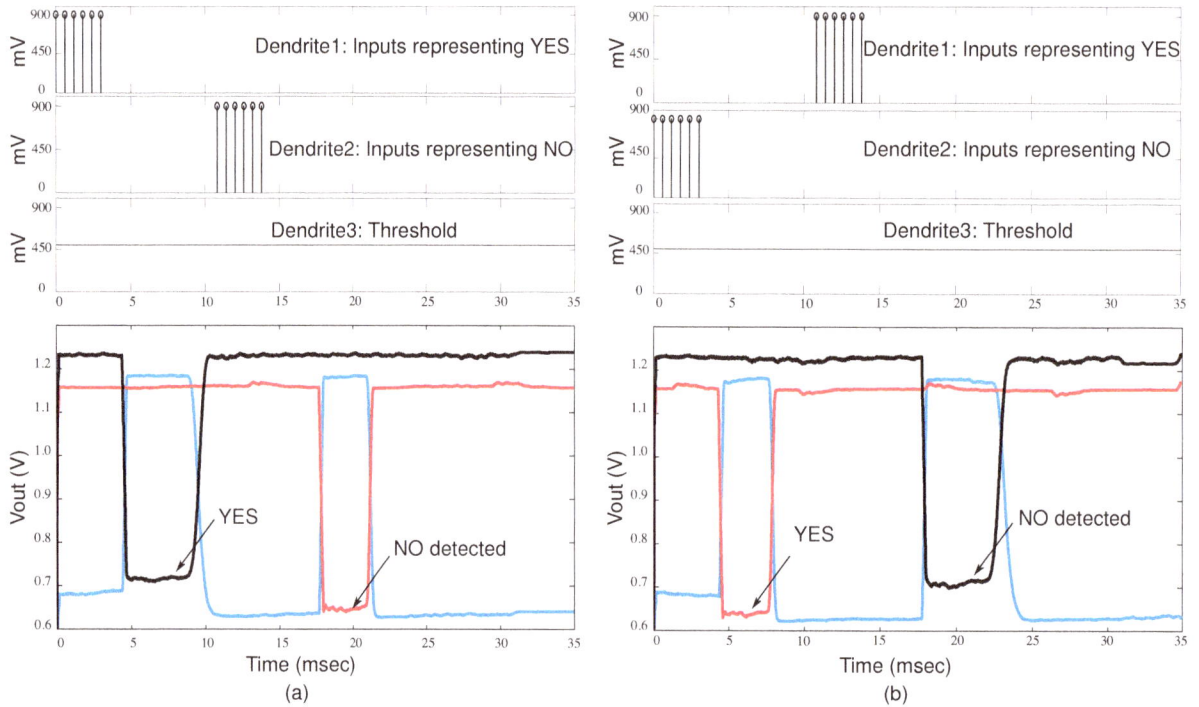

(a)   (b)

Figure 14. Experimental results for the classifier system with different timing delays for inputs and varying synaptic strength. It demonstrates the effect of different timing in sequences. This implementation is for a classifier system with 4-tap dendrites, 3-input WTA and supporting circuitry. The WTA outputs for varying delays between inputs is shown. The delay between subsequent inputs is $1.5\,ms$, $3\,ms$ and $9\,ms$ respectively. Also the inputs have different EPSP current strengths.

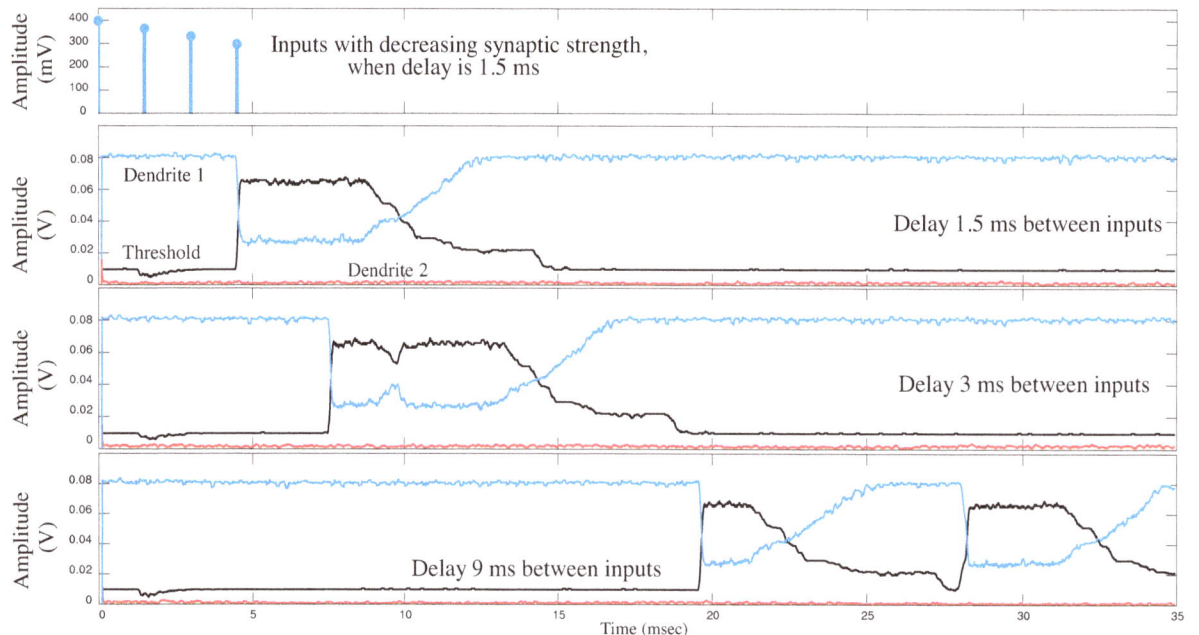

The winning output of the WTA is akin to an action potential. In terms of classification too, the WTA output signifies if a "word" has been detected. Our results have demonstrated that, such a system looks similar to an HMM state machine for a word/pattern. We can postulate from these experimental results that there are some similarities in computation done by HMM networks and a network of dendrites. The results are shown in Figure 12 for a single word and for continuous detection of words in Figure 13. We have demonstrated a biological model, built using circuits that is much closer than the implementation of any HMM network to date. Thus we have shown that an HMM classifier is possible using dendrites, and we have made a clearly neuromorphic connection to computation, a computation more rich than previously expected by dendritic structures.

5. Reconfigurable Platform to Build Neuromorphic Circuits

In the sections below, we will give a brief overview of the experimental setup used for the study. We used the FPAA, RASP $2.8a$ for all experimental data and the software tool MATLAB Simulink and $sim2spice$ script to build the dendrite simulation block.

5.1. FPAA Review

All the data presented in this paper comes from a reconfigurable hardware platform. The Field-Programmable Analog Array (FPAA) is a mixed-signal CMOS chip which allows analog components to be connected together in an arbitrary fashion. Reconfigurable Analog Signal Processor (RASP) was one of the first large scale FPAAs. It allowed us to build multiple complex circuits. The specific chip used from the family of RASP chips for this research work is RASP 2.8a [19]. It is a powerful and reconfigurable analog computing platform that can be used to build neuromorphic models. It consists of thirty-two Computational Analog Blocks (CABs). The CAB consists of groups of analog elements which include nFETs, pFETs, Operational Transconductance Amplifiers, capacitors, Gilbert multipliers, among others. These act as the computational elements which together can form complex sub-circuits that can be used to build analog computational systems. The interconnection of the CAB components is achieved by the switch matrix. It essentially consists of floating-gate (FG) pFETs. These $50,000$ programmable elements can be used not only as programmable interconnects for routing but also as adaptive computational elements. The switch matrix allows for both local routing between CAB elements as well as global routing. Last but not the least, it has the programmer block, which selectively accesses a floating-gate device on the chip and through tunneling and injection tune it on, off or operational in between. This is not only an efficient routing scheme but can enable implementation of dense systems.

5.2. Dendrite on the Routing Fabric

We used floating gate pFET switches to build the network of dendrites. This would also enable us to build denser networks as we scale the system. In our current system implementation for a single dendrite, we implemented 5 dendritic compartments, with each compartment consisting of 3 floating gate transistors. The most exciting aspect of implementing dendritic circuits using floating-gates is, that we can do so in a very compact manner. As stated above, the switch matrix of the RASP 2.8a FPAA is completely made up of about 50,000 floating-gate elements. Thus huge arrays of dendrites can be made using the switch matrix. Its inherent function is to interconnect components, which is similar to the function of dendrites that are used to transmit signals from one structure to another. Modeling dendritic circuits using floating gates, however has a few complications. The reason being the capacitive coupling from source and drain to the floating gate is more pronounced than regular pFETs [4]. Characterizing this capacitive coupling between the source and the drain is important if precision is desired. Another non-ideality that arises due to indirect programming is the mismatch between the transistor that is "programmed" *versus* the transistor that is actually used in the circuit. However, recently methods have been developed to characterize this mismatch [20].

Nevertheless, floating-gates enable building very compact circuits. This enables the building of larger systems like HMM classifiers using CMOS dendrites. The advantage being that not only could we individually program the FG-FETs for varying levels of charge to obtain taper easily but also could build a denser network. This would be useful for building larger systems. Also one must also take into account that neural systems are known to be inherently imprecise. Dendritic structures are not always similar and synapses are very unreliable. So one can say that this floating-gate mismatch is similar to dendrite-to-dendrite variability [4].

5.3. Simulink Model for Simulating CMOS Dendrites

Engineers have conventionally relied on digital systems like DSPs and FPGAs to implement algorithms for signal processing. A lot of software tools are available that enable and simplify this process. Thus existence of such intuitive software tools enables engineers to leverage the higher computational efficiency offered by hardware systems.Our lab has developed sim2spice, which is a tool that automatically converts analog signal processing systems from Simulink designs to a SPICE netlist [21]. It is the top-level tool in a complete chain of automation tools.The basic analog elements consist of the CAB elements on the FPAA. All parameters of the block are configurable. The Simulink block mainly serves two purposes. First, it converts the block-level Simulink model into a spice netlist which can be implemented on the FPAA. Secondly, it can also be used to run a behavioral simulation of the circuit.

5.3.1. Dendrite Simulink Block

The Simulink block simulates the behavioral characteristics of the dendrite structure given input/s. This provides the user an insight to the working of the dendritic circuit when implemented using the FPAA. The MOSFET parameters used are based on the MOSFETS present on the FPAA. It is characterized by coupled ordinary differential equations (ODE) and solved using the ode solver ode-45. The model has been tested for both static as well as time-varying inputs and has given reasonable results. For this paper we have used EPSP signals as inputs for the block. Consider a dendritic line as given in Figure 5c, with n number of nodes. The voltage at each node can be calculated using the following coupled ODE [4],

$$
\begin{aligned}
\frac{d\vec{V}}{dt} = \frac{1}{C}(a_1 \cdot I_{inj} &+ k_1(e^{a_2 \cdot \vec{V}/U_T} - e^{a_3 \cdot \vec{V}/U_T}) \\
&+ k_1(e^{a_4 \cdot \vec{V}/U_T} - e^{a_5 \cdot \vec{V}/U_T}) \\
&+ k_2(e^{a_6 \cdot \vec{V}/U_T} - e^{E_k/U_T}))
\end{aligned}
\tag{25}
$$

For taper, we changed the parameters k_1 as it is proportional to axial conductances.

6. Classifier: Computational Efficiency

Current approaches for Automatic Speech Recognition (ASR) use Hidden Markov Models as acoustic models for sub-word/word recognition and N-gram models for language models for words/word-class recognition. For HMMs, discriminative training methods have been found to perform better than other Maximum Likelihood methods like Baum-Welch estimation [22] . Our dendritic model is similar to a continuous-time HMM model and can be used to classify sub-phoneme, phonemes or words. Typically, phoneme recognition models have a much higher error rate as they are much less constrained as compared to word recognition models. Based on our comparison studies for different features we hypothesize that our model would have higher tolerance levels and dynamic range. We have not used an audio-dataset to characterize our system, rather we have used symbolic representations to make a hypothesis. These are experiments we plan to do int he near future. However, we can compare the computational efficiency of these methods since we can model these systems mathematically. The unit used to compare computational efficiency is Multiply ACcumulates (MAC) per Watt. The energy efficiency at a given node of the system, depends on the bias currents, supply voltage and also the node capacitance.

We know that the node capacitance C is the product of conductance and the time constant τ. Now the bias current I_{bias} for a dendrite node is given by,

$$
I_{bias} = (V_{rest} - E_k)\frac{C}{\tau}
\tag{26}
$$

where, V_{rest} is the resting potential; E_k signifies the voltage of a potassium channel and G is the axial conductance. Also, power is the product of voltage across the node and current into the node. Now for a single node of an HMM classifier, we have 2 MAC/sample. Assuming $\tau \sim delay$, which at a given node is approximately $1ms$. Thus,

$$
Energy/MAC = \frac{1}{2}V_{dd}(V_{rest} - E_k)C
\tag{27}
$$

We have compared the computational efficiency of digital, analog and biological systems as shown in Table 2. Now for a wordspotting passive dendritic structure, we have 2 MAC/node. Typical dendrite would have over 1000 state variable equivalents in its continuous structure. For a particular neuron time constant τ, we would want to have multiple samples for proper operation. For this discussion, let's assume an effective discrete time sample rate 5 times more than τ. Let us choose $\tau = 1ms$ for this discussion. Thus, we have each tree computing 10 MMAC for an HMM computation. For biological systems, say the brain has $1T$ neurons and total power consumption of about 20 W. Thus the power consumption is 20 pW/neuron. In a passive dendritic structure, the computational efficiency is 10 MMAC /neuron. Thus the computational efficiency of biological systems works out to be 0.5 MMAC/pW. Also from the equation it is evident that a major factor contributing to energy efficiency is node capacitance. Currently the node capacitance on the chip we used was $1pF$. If we further scale down the process used, this number will also reduce. This effectively means higher computational efficiency. A decrease to $10fF$ itself will give us an improvement of 2 orders of magnitude. This is depicted in Figure 15.

Table 2. Comparing computational efficiency of Digital, Analog and Biological systems.

Computing type	Computational efficiency
Digital (DSP)	< 10MMAC/mW [23]
Analog SP (VMM)	10MMAC/μW [11,24]
Analog (wordspotter)	> 10MMAC/μW
Neural process	> 10MMAC/pW

Figure 15. Computational efficiency *versus* capacitance plot for VMM (analog) and dendritic computation algorithms for $V_{dd} = 2.5V$ [25].

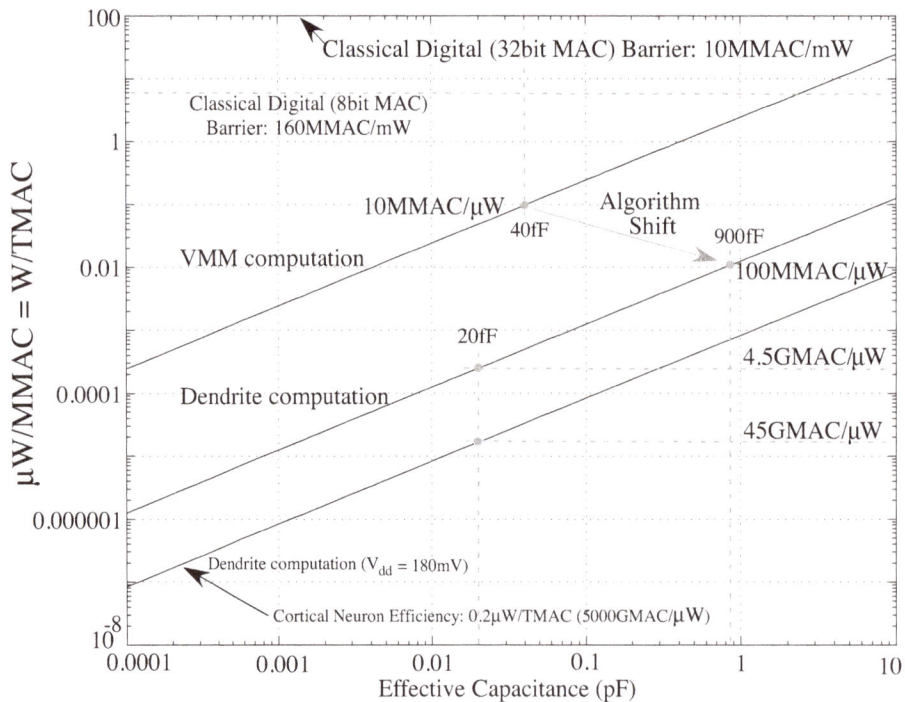

7. Conclusions

We have demonstrated a low-power dendritic computational classifier model to implement the state decoding block of a YES/NO wordspotter. We have also found that this implementation is computationally efficient. We have demonstrated a single dendritic line with 6 compartments, with each compartment having a single synaptic input current. We have seen the behavior of a single dendrite line by varying three parameters, namely, the "taper", the delay between inputs and the strength of the EPSP input currents. The effects of taper which enabled coincidence detection were studied. We have also seen the functioning of the WTA block with dendritic inputs and the how feedback helps initiate the reset after a word/phoneme is detected. We also build a Simulink dendritic model and simulated the output for time-varying inputs to compare with experimental data. This demonstrated how such a network would behave if inputs were in a sequence or if they were reversed.

The broader impact of such a system is two-fold. First, this system is an example of a computational model using bio-inspired circuits. Secondly the system proposes a computationally efficient solution for speech-recognition systems using analog VLSI systems. As we scale down the process, we can get more efficient and denser systems. We can also address how synaptic learning can be implemented and classification systems be trained. We can also model the input synapses as NMDA synapses to get a more multiplicative effect. In NMDA synapses, the synaptic strength is proportional to the membrane voltage. It couples the membrane potential to the cellular output. This could lead to a more robust system and is also closer to how biological systems are modeled. Also, we have modeled passive dendrites in this paper. It would be interesting to see how the system behaves when we add active channels. We currently have systems built that will enable us to further explore this discussion which is beyond the scope of this paper. There is a lot of scope for discussing how to build larger systems using this architecture. We can use spiking WTA networks for a larger dictionary of words. It is evident from the computational efficiency discussions, that clearly analog systems are a better choice for higher computational efficiency and lower costs. This calls for greater effort to build such systems. Reconfigurable/programmable analog systems open a wide range of possibilities in demonstrating biological processing and also for signal processing problems. As shown in Figure 16 there is great potential in other areas as image processing and communication networks as well. These systems will not only enhance our understanding of biological processes but also will help us design more computationally efficient systems.

Figure 16. Different applications using the Pattern Recognition system based on biology. It has application in speech and image processing and in communication systems. The state decoder in this paper is one block that is part of the whole system level design that we plan to build.

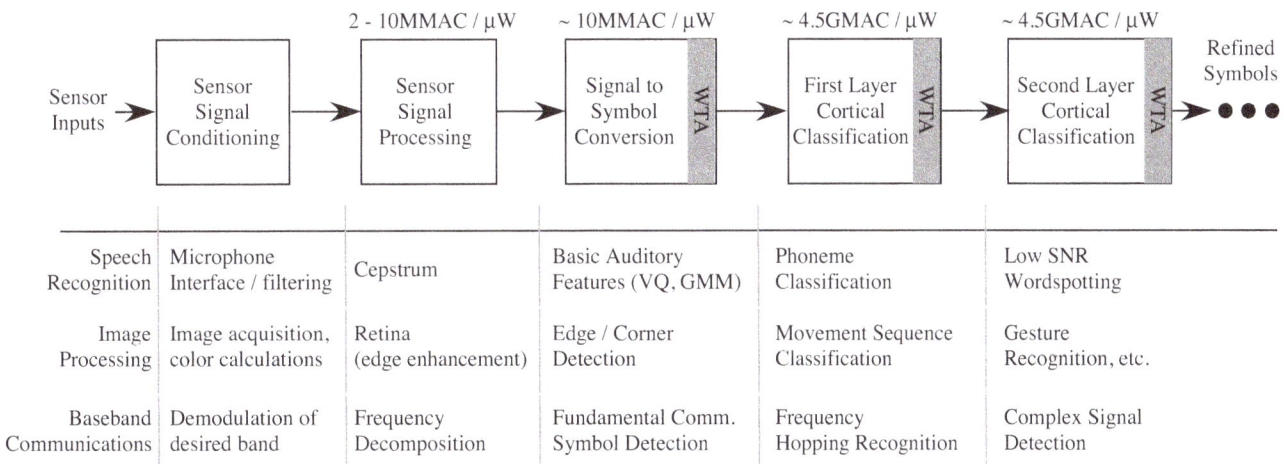

	2 - 10MMAC / µW	~ 10MMAC / µW	~ 4.5GMAC / µW	~ 4.5GMAC / µW	
Sensor Inputs →	Sensor Signal Conditioning →	Sensor Signal Processing →	Signal to Symbol Conversion / WTA →	First Layer Cortical Classification / WTA →	Second Layer Cortical Classification / WTA → ●●● Refined Symbols

Speech Recognition	Microphone Interface / filtering	Cepstrum	Basic Auditory Features (VQ, GMM)	Phoneme Classification	Low SNR Wordspotting
Image Processing	Image acquisition, color calculations	Retina (edge enhancement)	Edge / Corner Detection	Movement Sequence Classification	Gesture Recognition, etc.
Baseband Communications	Demodulation of desired band	Frequency Decomposition	Fundamental Comm. Symbol Detection	Frequency Hopping Recognition	Complex Signal Detection

Acknowledgments

We would like to sincerely thank Christopher Burdell, Tim Guglielmo, Carlos Solis and Ramona Diaz for their help with the initial classifier experiments.

References

1. Polsky, A.; Mel, B.W.; Schiller, J. Computational subunits in thin dendrites of pyramidal cells. *Nat. Neurosci.* **2004**, *7*, 621–627.

2. Wang, Y.; Liu, S.C. Input Evoked Nonlinearities in Silicon Dendritic Circuits. In Proceedings of the IEEE International Symposium on Circuits and Systems, Taipei, Taiwan, 24–27 May 2009; Volume 1, pp. 2894–2897.

3. Hasler, P.; Koziol, S.; Farquhar, E.; Basu, A. Transistor Channel Dendrites Implementing HMM Classifiers. In Proceedings of the IEEE International Symposium on Circuits and Systems (ISCAS), New Orleans, LA, USA, 27–30 May 2007; Volume 1, pp. 3359–3362.

4. Nease, S.; George, S.; Halser, P.; Koziol, S. Modeling and implementation of voltage-mode CMOS dendrites on a reconfigurable analog platform. *Biomed. Circuits Syst. IEEE Trans.* **2012**, *6*, 76–84.

5. Koch, C. *Biophysics of Computation*; Oxford University Press: New York, NY, USA, 1999.

6. London, M.; Hausser, M. Dendritic computation. *Ann. Rev. Neurosci.* **2005**, *28*, 503–532.

7. George, S.; Hasler, P. HMM Classifier Using Biophysically Based CMOS Dendrites for Wordspotting. In Proceedings of IEEE Biomedical Circuits and Systems Conference (BioCAS), San Diego, CA, USA, 10–12 November 2011; pp. 281–284.

8. Lippmann, R.P.; Chang, E.I.; Jankowski, C.R. Wordspotter Training Using Figure-of-Merit Back-Propagation. In Proceedings of International Conference on Acoustics, Speech, and Signal Processing, Adelaide, SA, USA, 19–22 April 1994; Volume 1, pp. 389–392.

9. Ramakrishnan, S.; Basu, A.; Chiu, L.K.; Hasler, P.; Anderson, D.; Brink, S. Speech Processing on a Reconfigurable Analog Platform. In Proceedings of the IEEE Subthreshold Microelectronics Conference (SubVT), Waltham, MA, USA, 9–10 October 2012; pp. 1–3.

10. Ramakrishnan, S.; Hasler, P. The VMM and WTA as an analog classifier. *IEEE Trans. VLSI Syst.* **2012**, in press.

11. Schlottmann, C.R.; Hasler, P.E. A highly dense, low power, programmable analog vector-matrix multiplier: The FPAA implementation. *IEEE J. Emerg. Sel. Top. Circuit Syst.* **2011**, *1*, 403–411.

12. Segev, I.; London, M. Untangling dendrites with quantitative models. *Science* **2000**, *290*, 744–750.

13. Lazzaro, J.; Wawrzynek, J.; Lippmann, R. A Micropower Analog VLSI HMM State Decoder for Wordspotting. In *Advances in Neural Information Processing Systems 9*; Mozer, M.C., Jordan, M.I., Petsche, T., Eds.; MIT Press: Cambridge, MA, USA, 1996; pp. 727–733.

14. Juang, B.H.; Rabiner, L.R. Hidden markov models for speech recognition. *Technometrics* **1991**, *33*, 251–272.

15. Hasler, P.; Smith, P.; Anderson, D.; Farquhar, E. A Neuromorphic IC Connection between Cortical Dendritic Processing and HMM Classification. In Proceedings of the IEEE 11th Digital Signal Processing and 2nd Signal Processing Education Workshop, Taos Ski Valley, NM, USA, 1–4 August 2004; pp. 334–337.

16. Mel, B.W. What the synapse tells the neuron. *Science* **2002**, *295*, 1845–1846.

17. Farquhar, E.; Abramson, D.; Hasler, P. A Reconfigurable Bidirectional Active 2 Dimensional Dendrite Model. In Proceedings of the IEEE International Symposium on Circuits and Systems, Vancouver, Canada, 23–26 May 2004; Volume 1, pp. 313–316.

18. Mead, C. *Analog VLSI and Neural Systems*; Addison-Wesley: Reading, MA, USA, 1989.

19. Basu, A.; Brink, S.; Schlottmann, C.; Ramakrishnan, S.; Petre, C.; Koziol, S.; Baskaya, F.; Twigg, C.; Hasler, P. A Floating-gate-based field programmable analog array. *IEEE J. Solid-State Circuits* **2010**, *45*, 1781–1794.

20. Shapero, S.; Hasler, P. Precise programming and mismatch compensation for low power analog computation on an FPAA. *IEEE Trans. Circuits Syst. I*, submitted for publication, 2013.

21. Schlottmann, C.R.; Petre, C.; Hasler, P.E. A high-level simulink-based tool for FPAA configuration. *IEEE Trans. VLSI Syst.* **2012**, *20*, 10–18.

22. Jiang, H.; Li, X.; Liu, C. Large margin hidden Markov models for speech recognition. *Audio Speech Lang. Process. IEEE Trans.* **2006**, *14*, 1584–1595.

23. Chawla, R.; Bandyopadhyay, A.; Srinivasan, V.; Hasler, P. A 531 nw/mhz, 128×32 Current-Mode Programmable Analog Vector-Matrix Multiplier with over Two Decades of Linearity. In Proceedings of the IEEE Conference on Custom Integrated Circuits, Orlando, FA, USA, 3–6 October 2004; pp. 29:1–29:4.

24. Marr, H.B.; Degnan, B.; Hasler, P.; Anderson, D. Scaling Energy per Operation via an Asynchronous Pipeline. *IEEE Trans. VLSI syst.* **2013**, *21*, 147–151.

25. Hasler, J.; Marr, B. Towards a roadmap for large-scale neuromorphic systems. *Front. Neurosci.* **2013**, Accept for publication.

3

Low-Power and Optimized VLSI Implementation of Compact Recursive Discrete Fourier Transform (RDFT) Processor for the Computations of DFT and Inverse Modified Cosine Transform (IMDCT) in a Digital Radio Mondiale (DRM) and DRM$^+$ Receiver

Shin-Chi Lai, Yueh-Shu Lee and Sheau-Fang Lei *

Department of Electrical Engineering, National Cheng Kung University, 701 Tainan, Taiwan;
E-Mails: chingivan2008@gmail.com (S.-C.L.); n28991201@mail.ncku.edu.tw (Y.-S.L.)

*   Author to whom correspondence should be addressed; E-Mail: leisf@mail.ncku.edu.tw

Abstract: This paper presents a compact structure of recursive discrete Fourier transform (RDFT) with prime factor (PF) and common factor (CF) algorithms to calculate variable-length DFT coefficients. Low-power optimizations in VLSI implementation are applied to the proposed RDFT design. In the algorithm, for 256-point DFT computation, the results show that the proposed method greatly reduces the number of multiplications/additions/computational cycles by 97.40/94.31/46.50% compared to a recent approach. In chip realization, the core size and chip size are, respectively, 0.84×0.84 and 1.38×1.38 mm^2. The power consumption for the 288- and 256-point DFT computations are, respectively, 10.2 (or 0.1051) and 11.5 (or 0.1176) mW at 25 (or 0.273) MHz simulated by NanoSim. It would be more efficient and more suitable than previous works for DRM and DRM$^+$ applications.

Keywords: digital radio mondiale (DRM); discrete fourier transform (DFT); inverse modified cosine transform (IMDCT); recursive structure

1. Introduction

Recently, the rapid growth of multimedia and wireless communication technologies has enabled the integration of various audio coding standards in a multimedia platform for audio applications. Digital Radio Mondiale (DRM) [1] is a digital broadcasting system for radio frequencies of below 30 MHz. DRM Plus (DRM$^+$) is the technology extending the DRM system to the VHF bands up to 174 MHz. It is a new standard of the European Telecommunication Standards Institute (ETSI ES 201 980). DRM offers the possibility to use various audio codecs, such as High Efficiency Advanced Audio Coding (HE-AAC), MPEG-4 Code-excited linear prediction (CELP), MPEG-4 Harmonic Vector Excitation Coding (HVXC), and MPEG Surround, *etc.*, to their broadcasting system. The discrete Fourier transform (DFT) and inverse modified cosine transform (IMDCT) have been, respectively, applied to realize the coded orthogonal frequency division multiplexing (COFDM) and the synthesis filterbank of advanced audio coding (AAC) in DRM specification. In a DRM and DRM$^+$ receiver, the COFDM adopts the non-power-of-two and power-of-two DFT whose transform lengths are specified to 288, 256, 176, 112, and 27. For a HE-AAC decoder, it also requires 1920- and 240-point IMDCT computations.

Previously, the issue of a common architecture design of fast Fourier transform (FFT) and IMDCT has been developed in [2–5] for a digital audio broadcasting (DAB) system [6,7]. The specified transform lengths of both FFT and IMDCT are all power of two, and it is very suitable for parallel design to implement the common architecture of FFT and IMDCT [2–4] or FFT-based IMDCT [5]. However, it would be a great challenge to design a flexible FFT or IMDCT accelerator with the transform lengths of power-of-two and non-power-of-two. Currently, Lai *et al.* [8] propose a recursive DFT (RDFT) to compute IMDCT coefficients. Due to the nature of recursive structure, Lai *et al.*'s hardware accelerator can arbitrarily switch transform length between power-of-two and non-power-of-two without any extra hardware processing units. Additionally, the results indicate that it achieves lower computational complexity than recursive DCT-based designs [9–18]. Since the FFT (or RDFT) can be further used for computing the IMDCT, the transform length of IMDCT is shortened from N to $N/4$. It exactly reduces the iteration loops for recursively computing IMDCT coefficients; however, compared with Lei *et al.*'s IMDCT [19], this method, which adopted one-dimension (1-D) RDFT as a unified kernel, would not gain better performance. This implies that it still has a bottleneck on using a 1-D RDFT to compute the IMDCT coefficients.

In 2004, Wolkotte *et al.* [20] presented a detailed analysis for computational complexity in a DRM receiver and clearly showed that the DFT block took 50.51% of all computations. To meet the specification that requires non-power-of-two transform lengths, RDFT approaches [21–32] and recursive IMDCT approaches [9–18] have been suggested for area-efficient implementations. The major limitation of RDFT is on the issue of long computational cycle, *i.e.*, long processing time. To overcome this shortcoming, a 2-D RDFT structure design with prime factor algorithm has been presented in [30–32]. Compared with other RDFT approaches [21–26], Lai *et al.* [29–32] have a greater improvement in terms of computational complexity and latency. However, the computational complexity of 256-point RDFT based on Lai *et al.*'s algorithm [32] still takes more multiplications and additions, because the transform length only has one prime factor so that we only can adopt three 1-D RDFT hardware accelerators to simultaneously compute all 256-point DFT coefficients. Another issue that arises in [30,32] is that, in order to decompose the N-point RDFT into c- and m-point sub-RDFTs, a great number of register files would be inserted between the first-stage and second-stage RDFTs for buffering temporal data. It makes the chip size become much larger, and consumes much power on memory access. Recently, a FFT design consisted of memory, control unit, and various mixed-radix butterfly modules has been presented in [33]. Hsiao *et al.* merge prime factor and common factor concepts to realize the proposed accelerator, and propose an efficient address generator to avoid the memory access conflict. However, it still requires many radix-r processing units to support the FFT computations. Due to the nature of RDFT, we can apply the variable-length and low-cost advantages to save these processing units. Thus, high-performance RDFT architecture is proposed to enhance our previous works [29–32] in this paper.

The rest of this paper is organized as follows: Section 2 takes an overview for our previous works [29–32] first, and then proposes a new concept to integrate them by applying the common factor DFT (CF-DFT) algorithm. Section 3 demonstrates the compact architecture design of the proposed RDFT, and then Section 4 introduces the low-power optimizations of VLSI implementation for the proposed hardware accelerator. Section 5 compares and contrasts the differences in performance for various approaches. Finally, conclusions are outlined in Section 6.

2. The Proposed Compact RDFT with Prime Factor and Common Factor Algorithms

The N-point DFT formula is defined as Equation (1). According to the derivation of Lai *et al.* [29], it can be found that the transfer function $H(z)$ is obtained as Equation (2).

$$X[k] = \sum_{n=0}^{N-1} x[n] \times W_N^{nk} \tag{1}$$

$$H(z) = \frac{M(z)}{X(z)} = \frac{W_N^{-k}}{1 - W_N^{-k}z^{-1}} = \frac{\cos\theta(k) + j\sin\theta(k) - z^{-1}}{1 - z^{-1}(2\cos\theta(k) - z^{-1})} \tag{2}$$

Equation (2) can be easily mapped into a hardware accelerator. To reduce the number of multiplications, both coefficients of $\cos\theta(k)$ and $2\cos\theta(k)$ can be computed by one multiplication and a simple shift operation. To reduce the usage of multipliers in implementation, a multiplier-sharing scheme is proposed and applied in [23,24,26,29–32] for this recursive structure. Hence, the multiplication of cosine and sine can be computed by the same multiplier with one clock cycle delay in realization. By adopting the hardware-sharing scheme and register-shifting concept, the RDFT circuit of Lai *et al.* [29] can be further improved. Figure 1 shows the compact RDFT circuit. The detailed control rules of multiplexers are shown in Figure 1b.

Figure 1. Proposed compact architecture of recursive discrete Fourier transform (RDFT) module. (**a**) The computational circuit of X[0] and X[N/2]; (**b**) The computational circuit design of X[1] to X[N/2 − 1].

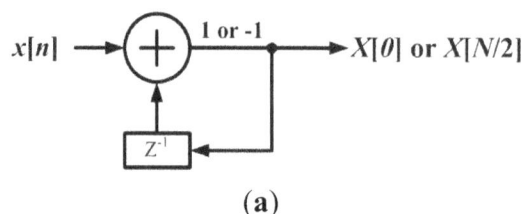

(a)

Figure 1. *Cont.*

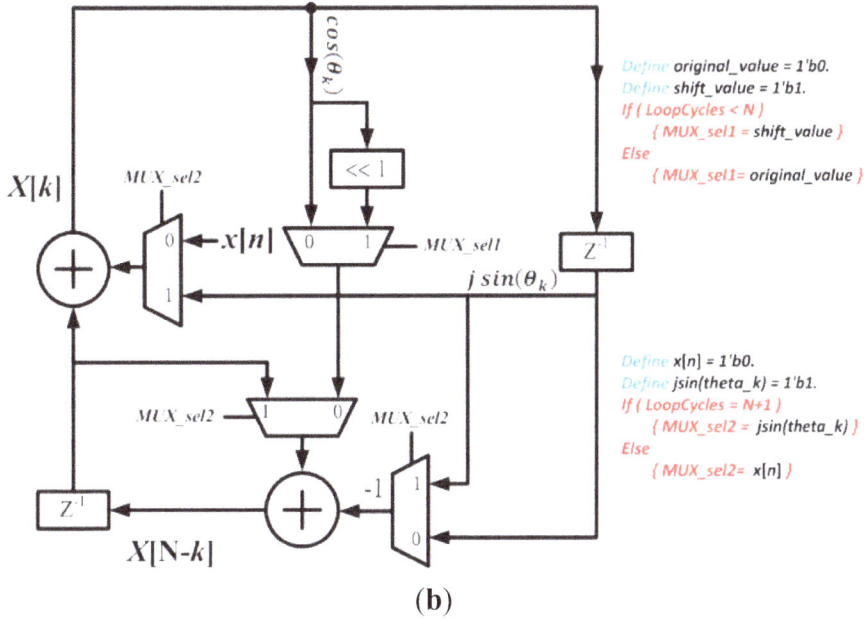

(b)

Compared with our previous work [29], this design only takes eight multiplexers, four adders and two multipliers in RDFT implementation. To lower the computational complexity and cycle, Lai *et al.* proposed a new algorithm using the Chinese Reminder Theorem (CRT), *i.e.*, prime factor algorithm, in [30–32]. In this algorithm, the input sequence length (N) can be factored into two mutually prime factors (c and m), and then the change of the variables is computed as Equation (3), where $\langle X \rangle_N = X \bmod N$.

$$n = \langle An_1 + Bn_2 \rangle_N; \; k = \langle Ck_1 + Dk_2 \rangle_N; \; n_1 \text{ and } k_1 = 0 \text{ to } c - 1; \; n_2 \text{ and } k_2 = 0 \text{ to } m - 1 \qquad (3)$$

For this mappings to be unique, the condition A, B, C, and D are chosen such that:

$$\langle AC \rangle_N = m \,, \langle BD \rangle_N = c, \text{ and } \langle AD \rangle_N = \langle BC \rangle_N = 0 \qquad (4)$$

Thus, the DFT algorithm with the CRT scheme, which is also called the prime factor DFT (PF-DFT) algorithm, can be defined as

$$X[\langle Ck_1 + Dk_2 \rangle_N] = \sum_{n_2=0}^{m-1} \left\{ \underbrace{\underbrace{\sum_{n_1=0}^{c-1} x[\langle An_1 + Bn_2 \rangle_N] W_c^{n_1 k_1}}_{c-point\ DFT}} W_m^{n_2 k_2} \right\} \qquad (5)$$

$$\underbrace{\phantom{X[\langle Ck_1 + Dk_2 \rangle_N] = \sum_{n_2=0}^{m-1} \left\{ \sum_{n_1=0}^{c-1} x[\langle An_1 + Bn_2 \rangle_N] W_c^{n_1 k_1} \right\} W_m^{n_2 k_2}}}_{m-point\ DFT}$$

By adopting the RDFT in Figure 1 to the Equation (5), the low-cost and reconfigurable two-dimension RDFT algorithm can be easily obtained. For a DRM system, the transform length of 288 (N) can be divided into 32 (c) and 9 (m) where factors of c and m are co-prime. The conditions of (A, B, C, D) are then chosen as (9, 32, 64, 225). Similarly, the conditions of (A, B, C, D) for the frame size of 176, 112, 480, 60 can be chosen as (11, 16, 33, 144), (7, 16, 49, 64), (15, 32, 225, 256), and (5, 12, 25, 36), respectively. Note that the details for the selection of these conditions are introduced in the section III of Lai *et al.* [30]. However, the PF-DFT algorithm cannot be applied to compute the 256-point DFT coefficients, because the transform length only has one prime, *i.e.*, $256 = 2^8$. Thus, we should adopt an efficient method called the CF-DFT algorithm to solve this problem. Assume that the input sequence length (N) can be also factored into two common factors (c and m), and

$$n = n_1 + mn_2, \begin{cases} n_1 = 0, \dots, m-1 \\ n_2 = 0, \dots, c-1 \end{cases}; k = ck_1 + k_2, \begin{cases} k_1 = 0, \dots, m-1 \\ k_2 = 0, \dots, c-1 \end{cases} \tag{6}$$

By taking Equation (6) into Equation (1), we can obtain the CF-DFT algorithm as Equation (7):

$$X[ck_1 + k_2] = \sum_{n_2=0}^{m-1} \left\{ \underbrace{\sum_{n_1=0}^{c-1} x[n_1 + mn_2] W_c^{n_1 k_1}}_{c-point\ DFT} W_N^{k_1 n_2} \right\} W_m^{n_2 k_2} \tag{7}$$

$$\underbrace{\phantom{X[ck_1 + k_2] = \sum_{n_2=0}^{m-1} \left\{ \sum_{n_1=0}^{c-1} x[n_1 + mn_2] W_c^{n_1 k_1} W_N^{k_1 n_2} \right\} W_m^{n_2 k_2}}}_{m-point\ DFT}$$

The difference between Equations (5) and (7) is the twiddle factor $W_N^{k_1 n_2}$. It implies that the computation between c-point DFT and m-point DFT has a complex multiplication. Additionally, it requires extra adders, multipliers and ROMs for the operations of twiddle factors in implementation. On the other hand, it will also increase the number of multiplications and additions in algorithm. Although the CF-DFT algorithm would take a fewer costs than PF-DFT, it can be applied to compute the one-prime-length DFT coefficients. To reduce the growth of coefficients with the variable-length DFT for DRM specification, Lai *et al.* proposes a coefficient-free algorithm in [31,32]. The two major coefficients in Equation (2), *i.e.*, $\cos\theta(k)$ and $\sin\theta(k)$, can be respectively calculated by using the trigonometric identities Equations (9,10) [31]. The detailed computations have been introduced in section 2.C of Lai *et al.*'s paper [31]. It is also applied to generate the coefficients of twiddle factors in this paper. Note that it only takes two computational cycles to generate the twiddle factors by using two multipliers in one RDFT kernel.

3. The Proposed Compact RDFT Architecture Design

In the previous section, a compact and low-cost RDFT circuit has been presented in detail. Due to the nature of RDFT, we can apply this design to implement PF-DFT and CF-DFT with variable transform-length DFT computations. Figure 2 demonstrates that the proposed compact RDFT architecture design. The proposed DFT processor can be briefly composed of two memory units, one controller unit, one RDFT unit, and some multiplexers.

Figure 2. Proposed compact RDFT architecture design.

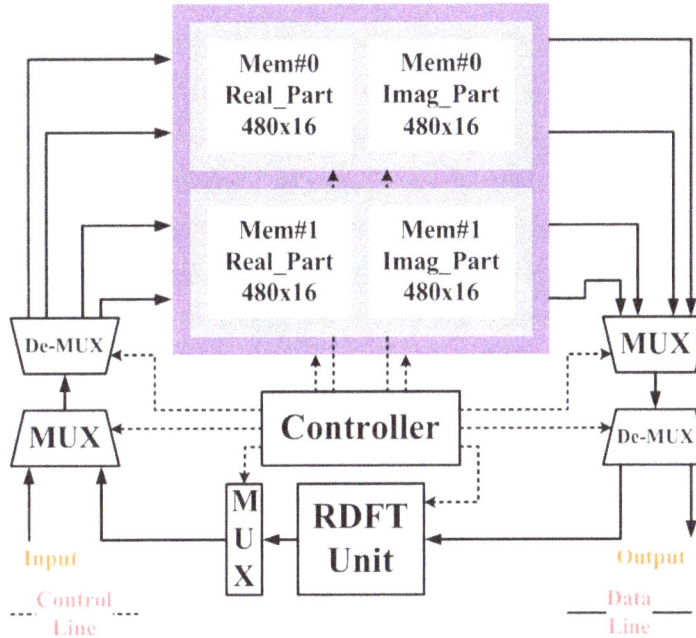

The time-domain sequence can be fed into the Mem#0 through the 32-bit input port with real and image parts. The data transmission time is corresponding to the transform length of DFT, and it takes N clock cycles for data transmission. While the input data storing in Mem#0 is ready, the c-point DFT computation can be started. First, the controller unit generates the address for the memory to access the required data, and then the read-out data are further fed into the RDFT unit. Due to the characteristic of the proposed RDFT circuit, two coefficients, *i.e.*, $X[k]$ and $X[N-k]$, can be obtained, and thus we require two 32-bit single port RAMs (16-bit for real part and 16-bit for image part) to alternately switch for storing the temporal data. While the RDFT figures the results out, the controller unit would send a signal for MUX unit to make these two parallel DFT coefficients be sequentially stored back to the Mem#1. After it finishes m times c-point DFT computations, the Mem#1 will fully store all data for the further m-point DFT computation. Hence, we can repeat the above computational flow to generate the m-point DFT coefficients, and then fed the results into Mem#0. Finally, the controller unit will send a signal to the multiplexer and makes the all frequency-domain sequence be transferred to 32-bit output port through the Mem#0. It will also take N clock cycles for data transmission. Note that the twiddle factor multiplications are computed by the same RDFT module, if the CF-DFT algorithm is adopted.

4. Low Power VLSI Implementation

In this section, we will introduce some low-power and optimized schemes applied for the proposed design and further demonstrate the implementation results in detail.

4.1. Low-Power Optimizations

The power consumption has become a critical issue for VLSI design in recent years. For most designs, all designers try their best to minimize the power consumption of chip by using some low-power optimizations such as (1) Clock gating; (2) Operand isolation; and (3) Voltage scaling. For clock gating scheme, it is a method that often used in low power designs, and provides a way to selectively close the clock. It forces the original circuit to make no transition whenever the computation carrying out at the next clock cycle is redundant. In other words, the clock signal is disabled according to the idle conditions of the logic network. We use this gated clock mechanism on the memory module since the memories need to access frequently and it may cause higher power consumption. However, in our design, the number of memory access is processed in an interleaved manner, and the RDFT module needs c or m cycles for each computation. It implies that the storage data will be not updated every cycle, and we can therefore close the clock to reduce power consumption. The idea of operand isolation is to identify redundant operations and uses special isolation circuitry to prevent switching activity from propagating into a module whenever it is to perform a redundant operation [34]. Therefore, the transition activity of the internal nodes of the modules can be significantly reduced, and thus it has lower power consumption. However, the operand isolation technique needs to add extra logic, and may cause some overhead for systems. In the proposed design, the power consumption is dominated on the memory and the RDFT modules. The inputs of RDFT module is coming from memory so that we can isolate the memory output by using a chip enable pin. This method is good for operand isolation since we only need to add some combination circuit to control the enable signal. For the voltage scaling scheme, the traditional dynamic power dissipation equation is defined as

$$P = \alpha \times f \times C \times V^2 \tag{8}$$

where α is the switching activity, f is the operating frequency, C is loading capacitance, and V is the supply voltage. The supply voltage affects dynamic power by a square term so that to decrease power supply on designs is an efficient way to reduce the power dissipation. Figure 3 shows the power dissipation of the proposed design for 288-point DFT computations at different supply voltages. It clearly shows that the voltage scaling technique can decrease the power consumption by 29.5%.

Figure 3. Power dissipation of 288-point DFT computation at different supply voltages for the proposed RDFT design.

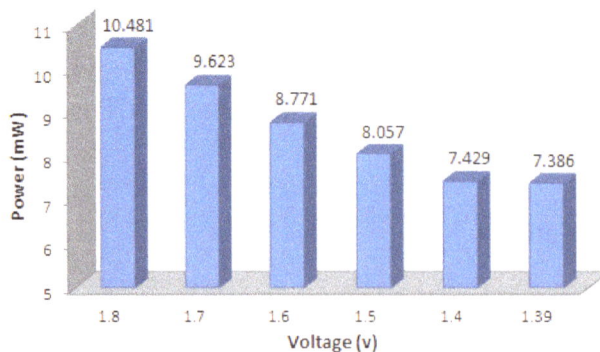

4.2. Implementation Results

The proposed RDFT processor is implemented by using the cell-based design flow with the TSMC 0.18 μm 1P6M CMOS technology. The input and output word lengths are both set to 16-bit format. The internal and coefficient word lengths are both set to 24 bits. Each memory block, *i.e.*, a register file, is generated by Artisan's Memory Compiler. The verilog code is simulated using Verilog-XL and then is synthesized using the Design Compiler. Finally, it is floorplaned for layout using SoC-encounter. According to the simulation result, the number of gate count of Memory/RDFT/Controller/MUXs is 31,332/13,594/13,451/1,218. Figure 4 shows the percentage of gate count for each module in the proposed design. Figure 5 demonstrates the layout of the proposed RDFT chip.

Figure 4. Percentage of gate count for the proposed design.

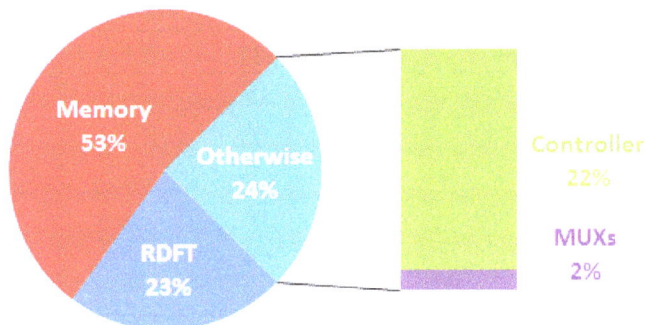

The core area and chip area are 0.85×0.84 and 1.38×1.37 mm^2, respectively. The power consumption for 288- and 256-point computation without adopting voltage scaling scheme are, respectively, 10.48 mW and 11.44 mW @25MHz simulated by Nanosim. In fact, the proposed RDFT processor can operate at 273 kHz to meet the real-time requirement of DRM standard. It implies that the power dissipation of the proposed design can be, respectively reduced to 105.1 and 117.6 μW for 288- and 256-point computation. Since the proposed design employs the voltage scaling scheme, the power consumption of the proposed design is reduced from 10.48 mW down to 7.39 mW.

Figure 5. Chip layout of the proposed RDFT processor.

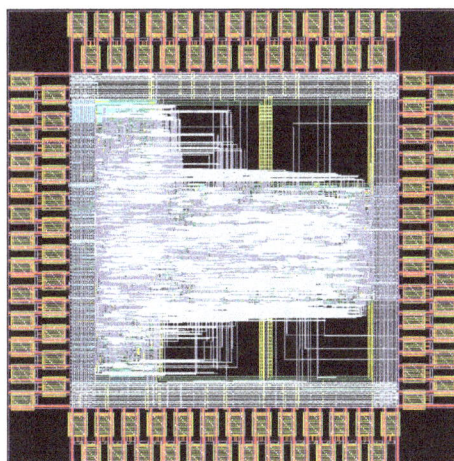

5. Comparison and Discussion

In this section, we make a completed comparison for various performance evaluations in terms of computational complexity, computational cycle, time cost per transformation (TCPT), and hardware costs. Since the proposed method adopted CF and PF algorithms to speed up the traditional RDFT computation, the factor selection would be greatly impact on the performance of the proposed RDFT.

Here, we compare all lists of the required DFT transform lengths in the OFDM of DRM standard, *i.e.*, 288, 256, 176, 112, and 27. Table 1 lists the suitable *c* and *m* factors for the proposed RDFT. Both *c* and *m* factors are corresponding to the number of computational complexity and the number of computational cycle. To avoid the twiddle factor multiplications in the proposed hardware design, the transform lengths of 288, 176, and 112 are calculated by PFA. Only 256- and 27-point DFT are computed by CFA.

Table 1. Factor selection for the proposed RDFT.

Length	288	256 *	176	112	27 *	480	60
c	32	16	16	16	9	32	12
m	9	16	11	7	3	15	5
A	9	1	11	7	1	15	5
B	32	*m*	16	16	*m*	32	12
C	64	*c*	144	64	*c*	256	36
D	255	1	33	49	1	255	25

Note: * common factor only.

Table 2 demonstrates a comparison of computational complexity for various RDFTs with different transform lengths. Van *et al.*'s method [26] takes $(2N^2 + 6N)$ multiplications and $(4N^2 + 8N)$ additions to compute N-point DFT coefficients. On the other hand, Lai *et al.* [9] proposes a much simpler structure, and it takes $(N^2 - N - 2)$ multiplications and $(2N^2 + 7N - 2)$ additions in *N*-point DFT computation. Furthermore, Lai *et al.* proposed two low-complexity methods [30,32] to reduce the numbers of multiplication and addition. It only takes $[2N(m + c + 2)]$ multiplications and $[4N(m + c + 2)]$ additions in [30]. For the proposed PF-RDFT algorithm, *i.e.*, in case of 288-point DFT computations, the number of multiplications would, respectively, take $[2m(c + 1)(c/2 - 1)]$ and $[2c(m + 1)(m - 1)/2]$ for *c*-point and *m*-point RDFT computations as shown in (5), where $(c/2 - 1)$ and $[(m - 1)/2]$ are the corresponding number of recursive loops, respectively, for even and odd-point RDFT. In addition, $[2m(c + 1)]$ and $[2c(m + 1)]$ are respectively the number of multiplications per recursive loop, where the scale "2" means the multiplication required for complex input sequence. The total number of additions would, respectively, take $[4N(c/2 - 1) + 4c]$ and $[4N(m - 1)/2 + 4m]$ for *c*-point and *m*-point RDFT computations, where $(4N)$ is the number of additions per recursive loop in Figure 1b, and $(4c)$ is the total number of additions only for Figure 1a under consideration of the case of *c*-point RDFT computations. About the proposed CF-RDFT algorithm, *i.e.*, in case of 256-point DFT computations as shown in Equation (7), it requires extra multiplications and additions for the twiddle factor operation more than that of the proposed PF-RDFT algorithm. However, the twiddle factor operation only takes $(4N)$ multiplications and $(2N)$ additions.

Since the proposed method has a compact and high-throughput RDFT circuit as well as our previous work [32], we can further calculate the desired coefficients with much lower complexity by combining PF and CF algorithms with RDFT. The result shows that the proposed method can, respectively, reduce the numbers of multiplications and additions by 97.40% and 94.31% for 256-point DFT computation while the CFA is adopted.

Table 2. Computational complexity of various RDFT algorithms for 288- and 256-point computations.

Method	Multiplications		Additions	
	N = 288	N = 256	N = 288	N = 256
[26] *	167,616	132,608	334,080	264,192
[27]	41,464	32,760	85,658	67,946
[29]	82,654	65,278	167,902	132,862
[30]	24,768	332,928	49,536	263,168
[32]	12,704	332,928	25,984	263,168
Proposed	11,470	8,640	22,034	14,976

Note: * pre-processing excluded.

To evaluate the latency of various algorithms, we make a comparison of computational cycle with five transform lengths specified in DRM system. Van *et al.* [26] and Lai *et al.* [29] require ($N^2/2$) and $[(N^2 - N - 1)/2]$ cycles, respectively, to compute the N-point DFT coefficients. In Table 3, Meher *et al.* [27] has the lowest computational cycle but requires most hardware resources to implement the systolic structure. Since PFA can be applied to speed up the conventional RDFT and makes the N-point DFT to be c- and m-point sub-DFTs, Lai *et al.* [30,32] have a lower latency in computation. For example, Lai *et al.* [30] only requires $[(c + 1)N + (m + 1)m]$ cycles to calculate all coefficients for 288, 176, and 112 transform lengths. Due to the difference of RDFT kernel design, the result shows that the number of computational cycle for our previous works [30,32] is relatively different. In this work, we propose a memory-based structure with a single and compact RDFT processing unit to compute variable-length transform. The results show that the proposed method has a lower latency compared with previous works [26,29,30]. Compared to [32], although the proposed method requires more latency for 288-, 176-, 112-, and 27-point DFT, it still shows a greater performance in terms of 256-point DFT computation, and dramatically reduces the computational cycle by 46.50%. In addition, the hardware costs of this work can be greatly improved, as shown in Table 4.

Table 3. Computational cycle of various RDFT algorithms.

Method	Transform length				
	288	256	176	112	27
[26]	41,472	32,768	15,488	6,272	N/A
[27]	431	383	263	167	N/A
[29]	41,327	32,639	15,399	6,215	364
[30]	9,594	32,896	3,124	1,960	N/A
[32]	4,842	11,137	1,594	1,006	N/A
Proposed	6,693	5,958	2,865	1,609	346

For hardware cost comparison, we use items such as a multiplier, adder, buffer, coefficient-ROM, and data throughput per transformation (DTPT) to evaluate the existing RDFT designs in Table 4. For the temporal buffer, Meher *et al.* [27] require ($N/4 - 2$) latch cells and each latch cell takes four latches to store the data. By the way, Lai *et al.* [30] and [32], respectively, require (4×11) and (8×15)

register files. On the other hand, for the number of coefficients stored in the ROM, Van *et al.* [26] and Lai *et al.* [29], respectively require (N) and ($N - 2$) coefficients for each N-point DFT computation, but Meher *et al.* [27] require ($3N - 2$) coefficients. It can be observed that Lai *et al.* [32] have more hardware costs than previous works [26,29,30] do; however, the complexity and latency of [32] have a better improvement according to Tables 2 and 3. To make the design balance in realization, *i.e.*, cost and performance, the proposed method adopts only one RDFT module to implement the memory-based DFT processor. Thus, it does not require extra buffer for temporal data as is the case with Meher *et al.* [27] and Lai *et al.* [30,32]. Additionally, we adopt on-line coefficient generator [31,32] to avoid coefficient-ROM using in chip implementation and maintain the same DTPT as well as [29]. The overall comparisons and considerations are clearly indicated that the proposed solution would be a low-cost and high-performance design for variable-length DFT computations.

Table 4. Hardware cost for various RDFT algorithms.

Method	Multiplier	Adder	Buffer	ROM	DTPT
[26]	10	17	No	Yes	1
[27]	N + 4	N + 18	Yes	Yes	4
[29]	2	13	No	Yes	2
[30]	4	8	Yes	Yes	1
[32]	6	18	Yes	No	4
Proposed	2	4	No	No	2

The time cost per transformation (TCPT) of various-length DFTs specified by DRM standard is summarized in Table 5. Based on the results of Table 3, the TCPT of the proposed design can be estimated while the operating frequency rate is set to 25 MHz. In addition, the RAM access time per transformation (RAM_ATPT) is used to calculate the data loading into DFT accelerator and storing back to the system bus, since the proposed design would be an IP in a system. The results show that it can easily achieve real-time requirement of DRM standard and saves over 98.91% of time cost. It implies that we can adjust the operation frequency down to 273 kHz to achieve the requirement of low power consumption. According to NanoSim simulation results, the power consumption for the 288- and 256-point DFT computations are, respectively, 0.105 mW and 0.1176 mW at 273 kHz.

Table 5. Time cost per transformation for the proposed method.

Length	288	256	176	112	27
DRM Spec. (ms)	<26.7	<26.7	<20	<16.7	<2.5
Proposed (us)	267.72	238.32	114.60	64.36	13.84
RAM_ATPT(us)	23.04	20.48	14.08	8.96	2.16
Reduction (%)	98.91	99.03	99.36	99.56	99.36

Table 6 summarizes the comparisons between the proposed design and other RDFT processors for DRM receiver in the literature. For the purpose of fair comparison with different process technologies, we employ Baas' normalization Equations (9) and (10) [35] to normalize the area and DFTs/Energy.

$$Normalized\ Area = \frac{Area}{(Technology/0.18\mu m)^2} \qquad (9)$$

$$\frac{DFTs}{Energy} = \frac{(Technology/0.18\mu m)(VDD/1.8)^2}{Power \times Execution\ Time \times 10^3} \tag{10}$$

Table 6. Comparison of chip feature for previous works and this work.

Design		[29]	[30]	[33]	This work
Technology		0.18 μm	0.18 μm	0.18 μm	0.18 μm
Internal/Coeff. word lengths		24/24 (bits)	21/16 (bits)	24/24 (bits)	24/24 (bits)
Data Memory (bits)		Excluded	Excluded	Excluded	$2 \times 480 \times 32$
Coefficient Memory		Excluded	Coeff.-free	Coeff.-free	Coeff.-free
Supply Voltage		1.98 v	1.98 v	1.98 v	1.7 v (opt.)
Clock Rate		25 MHz	25 MHz	25 MHz	25 MHz
Supporting DFT		288, 256, 176,	288, 256, 176,	288, 256, 176,	288, 256, 176,
Transform-Length		112, 212, 106	112	112, 480, 60	112, 480, 60
Executing Time for 288-point		1.65 ms	384 μs	193.68 μs	267.72 μs
Power	Circuit	5.98 mW	8.44 mW	14.3 mW	9.62 mW(opt.)
Consumption	Data Memory	5.53 mW *	5.53 mW*	5.53 mW *	
Core Area	Circuit	0.154 mm^2	0.265 mm^2	0.746 mm^2	0.714 mm^2
	Data Memory	0.347 mm^2	0.347 mm^2	0.347 mm^2	
Normalized DFTs/Energy		63.71	225.56	315.05	346.34 (opt.)

Note: * estimated by this work.

Note that the memory unit excluded in these literatures. Van *et al.*'s algorithm [26] has a number of $N^2/2$ computational cycles as well as Lai *et al.*'s [29], but the implementation chip of Van *et al.* is only designed to DTMF application. For DRM applications, previous works [30,32] have a better performance in terms of normalized DFTs per energy; however, these designs do not include the data memory to buffer the input and out sequence. This implies that these two designs would occupy most bandwidth of system bus, since they are hard intellectual properties (IPs) in an embedded system. Form the system view to consider this problem, the area and power consumption of data memory (RAM) should not be neglected. Table 2 clearly shows that the proposed design is a smaller than previous work [32], although it seems to lose its advantage in power consumption. Here, we also provide a power consumption result simulated by NanoSim. The result shows that the proposed design consumes 10.693 mW @ 25 MHz. The percentage of power dissipation is as shown in Figure 6. We can see that it consumes the most of power in RDFT module, and the power consumption of memory module is 1.925 mW. Based on Figures 4 and 6, we know that the memory module consumes approximately a fifth of total power, and takes more chip area over 50% of total gate counts.

Figure 6. Percentage of power dissipation for the proposed design by Prime Power.

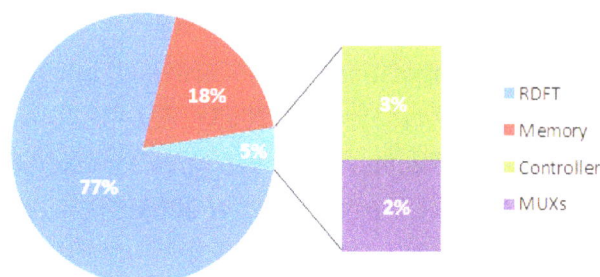

6. Conclusions

This paper presented a high-performance design for variable-length DFT computations for a DRM and DRM$^+$ receiver by using the low-power and optimized VLSI schemes in implementation. In addition, the compact RDFT kernel integrates prime factor and common factor algorithms into one structure, and only costs a smaller area than previous designs. Therefore, it would be a regular, flexible, and compact design for a VLSI realization in many future variable-length DFT and IMDCT computations.

Acknowledgments

This work was supported in part by the National Science Council, Taiwan under Grant No. 101-2218-E006-005 and 101-2221-E006-271.

References

1. *Digital Radio Mondiale*; *System Specification*; ES 201 980 V3.1.1; European Telecommunications Standards Institute (ETSI): Nice, France, August 2009.
2. Tai, S.C.; Wang, C.C.; Wang, J.L. Circuit-Sharing Design between FFT and IMDCT with Pipeline Structure for DAB Receiver. In Proceedings of the 17th International Conference on Advanced Information Networking and Applications, Xi'an, China, 27–29 March 2003; pp. 768–773.
3. Tai, S.C.; Wang, C.C.; Lin, C.Y. FFT and IMDCT circuit sharing in DAB receiver. *IEEE Trans. Broadcast.* **2003**, *49*, 124–131.
4. Wang, C.C.; Lin, C.Y. An Efficient FFT processor for DAB receiver using circuit-sharing pipeline design. *IEEE Trans. Broadcast.* **2007**, *53*, 670–677.
5. Kim, B.E.; Chung, J.Y.; Hwang, S.Y. An efficient fixed-point IMDCT algorithm for high-resolution audio appliances. *IEEE Trans. Consum. Electron.* **2008**, *54*, 1867–1872.
6. *Radio Broadcasting System*: *Digital Audio Broadcasting to Mobile Portable and Fixed Receiver*; ETS 300 401; European Telecommunications Standards Institute (ETSI): Nice, France, January 2006.
7. *Digital Audio Broadcasting (DAB)*: *Transport of Advanced Audio Coding (AAC) Audio*; ETSI TS 102 563; European Telecommunications Standards Institute (ETSI): Nice, France, February 2007.
8. Lai, S.C.; Lei, S.F.; Luo, C.H. Low-Cost and Shared Architecture Design of Recursive DFT/IDFT/IMDCT Algorithms for Digital Radio Mondiale System. In Proceedings of IEEE International Conference on Intelligent Information Hiding and Multimedia Signal Processing (IIH-MSP-2010), Darmstadt, Germany, 15–17 October 2010; pp. 276–279.
9. Chiang, H.C.; Liu, J.C. Regressive implementations for the forward and inverse MDCT in MPEG audio coding. *IEEE Signal Process. Lett.* **1996**, *3*, 116–118.
10. Nikolajevic, V.; Fettweis, G. Computation of forward and inverse MDCT using Clenshaw's recurrence formula. *IEEE Trans. Signal Process.* **2003**, *51*, 1439–1444.
11. Chen, C.G.; Liu, B.D.; Yang, J.F. Recursive architectures for realizing modified discrete cosine transform and its inverse. *IEEE Trans. Circuits Syst. II* **2003**, *50*, 28–45.

12. Nikolajevič, V.; Fettweis, G. New Recursive Algorithms for the Forward and Inverse MDCT. In Proceedings of the IEEE Workshop on Signal Processing Systems: Design and Implementation (SiPS'2001), Antwerp, Belgium, 26–28 September 2001; pp. 51–57.

13. Nikolajevič, V.; Fettweis, G. New recursive algorithms for the unified forward and inverse MDCT/MDST. *J. VLSI Signal Process. Syst.* **2003**, *34*, 203–208.

14. Fox, W.; Carriera, A. Goertzel Implementations of the Forward and Inverse Modified Discrete Cosine Transform. In Proceedings of the IEEE Canadian Conference on Electrical and Computer Engineering (CCECE'2004), Niagara Falls, Canada, 2–5 May 2004; pp. 2371–2374.

15. Chen, C.H.; Wu, C.B; Liu, B.D.; Yang, J.F. Recursive Architectures for the Forward and Inverse Modified Discrete Cosine Transform. In Proceedings of the IEEE Workshop on Signal Processing Systems: Design and Implementation (SiPS'2000), Lafayette, LA, USA, 11–13 October 2000; pp. 50–59.

16. Cheng, Z.Y.; Chen, C.H.; Liu, B.D.; Yang, J.F. Unified Selectable Fixed-Coefficient Recursive Structures for Computing DCT, IMDCT and Subband Synthesis Filtering. In Proceedings of the IEEE International Symposium on Circuits and Systems, Vancouver, Canada, 23–26 May 2004; pp. 557–560.

17. Lei, S.F.; Lai, S.C.; Hwang, Y.T.; Luo, C.H. A High-Precision Algorithm for the Forward and Inverse MDCT Using the Unified Recursive Architecture. In Proceedings of the IEEE International Symposium on Consumer Electronics, Vilamoura, Algarve, 14–16 April 2008; pp. 1–4.

18. Lai, S.C.; Lei, S.F.; Luo, C.H. Common architecture design of novel recursive MDCT and IMDCT algorithms for application to AAC, AAC in DRM, and MP3 codecs. *IEEE Trans. Circuits Syst. II* **2009**, *56*, 793–797.

19. Lei, S.F.; Lai, S.C.; Cheng, P.Y.; Luo, C.H. Low complexity and fast computation for recursive MDCT and IMDCT algorithms. *IEEE Trans. Circuits Syst. II* **2010**, *57*, 571–575.

20. Wolkotte, P.T.; Smit, G.J.M.; Smit, L.T. Partitioning of a DRM Receiver. In Proceedings of the 9th International OFDM-Workshop, Dresden, Germany, 15–16, September 2004; pp. 299–304.

21. Goertzel, G. An algorithm for the evaluation of finite trigonometric series. *Am. Math.* **1958**, *65*, 34–35.

22. Yang, J.F.; Chen, F.K. Recursive discrete Fourier transform with unified IIR filter stluclures. *Signal Process.* **2002**, *82*, 31–41.

23. Van, L.D.; Yang, C.C. High-Speed Area-Efficient Recursive DFT/IDFT Architectures. In Proceedings of the IEEE International Symposium Circuits and System, Vancouver, Canada, 23–26 May 2004; pp. 357–360.

24. Van, L.D.; Yu, Y.C.; Huang, C.N.; Lin, C.T. Low Computation Cycle and High Speed Recursive DFT/IDFT: VLSI Algorithm and Architecture. In Proceedings of the IEEE Workshop on Signal Processing Systems, Athens, Greece, 2–4 November 2005; pp. 579–584.

25. Fan, C.P.; Su, G.A. Efficient recursive discrete Fourier transform design with low round-off error. *Int. J. Electr. Eng.* **2006**, *13*, 9–20.

26. Van, L.D.; Lin, C.T.; Yu, Y.C. VLSI architecture for the low-computation cycle and power-efficient recursive DFT/IDFT design. *IEICE Trans. Fundam. Electron. Commun. Comput. Sci.* **2007**, *E90-A*, 1644–1652.

27. Meher, P.K.; Patra, J.C.; Vinod, A.P. Novel Recursive Solution for Area-Time Efficient Systolization of Discrete Fourier Transform. In Proceedings of the IEEE International Symposium on Signals, Circuits and Systems, Lasi, Romania, 12–13 July 2007; pp. 193–196.

28. Meher, P.K.; Patra, J.C.; Vinod, A.P. Efficient systolic designs for 1- and 2-dimensional DFT of general transform-lengths for high-speed wireless communication applications. *J. Signal Process. Syst.* **2010**, *60*, 1–14.

29. Lai, S.C.; Lei, S.F.; Chang, C.L.; Lin, C.C.; Luo, C.H. Low computational complexity, low power, and low area design for the implementation of recursive DFT and IDFT algorithms. *IEEE Trans. Circuits Syst. II* **2009**, *56*, 921–925.

30. Lai, S.C.; Juang, W.H.; Chang, C.L.; Lin, C.C.; Luo, C.H.; Lei, S.F. Low-computation cycle, power-efficient, and reconfigurable design of recursive DFT for portable digital radio mondiale receiver. *IEEE Trans. Circuits Syst. II* **2010**, *57*, 647–651.

31. Lai, S.C.; Lei, S.F.; Juang, W.H.; Luo, C.H. A low-cost, low-complexity and memory-free architecture of novel recursive DFT and IDFT algorithms for DTMF application. *IEEE Trans. Circuits Syst. II* **2010**, *57*, 711–715.

32. Lai, S.C.; Juang, W.H.; Lin, C.C.; Luo, C.H.; Lei, S.F. High-throughput, power-efficient, coefficient-free and reconfigurable green design for recursive DFT in a portable DRM receiver. *Int. J. Electr. Eng.* **2011**, *18*, 137–145.

33. Hsiao, C.F.; Chen Y.; Lee, C.Y. A generalized mixed-radix algorithm for memory-based FFT processors. *IEEE Trans. Circuits Syst. II* **2010**, *57*, 26–30.

34. Munch, M.; Wurth, B.; Mehra, R.; Sproch, J.; Wehnl, N. Automating RT-Level Operand Isolation to Minimize Power Consumption in Datapaths. In Proceedings of the IEEE Design Automation and Test, Paris, France, 27–30 March 2000; pp. 624–631.

35. Baas, B.M. A low-power, high-performance, 1024-Point FFT processor. *IEEE J. Solid-State Circuits* **1999**, *34*, 380–387.

4

Sub-Threshold Standard Cell Sizing Methodology and Library Comparison

Bo Liu [1,2,*]**, Jose Pineda de Gyvez** [1] **and Maryam Ashouei** [2]

[1] Electronic System Group, Department of Electrical Engineering, Technische Universiteit Eindhoven, Den Dolech 2, 5612AZ, Eindhoven, The Netherlands; E-Mail: J.Pineda.de.Gyvez@tue.nl

[2] Holst Centre/Imec-nl, High Tech Campus 31, 5656AE, Eindhoven, The Netherlands; E-Mail: Maryam.Ashouei@imec-nl.nl

* Author to whom correspondence should be addressed; E-Mail: B.liu@tue.nl

Abstract: Scaling the voltage to the sub-threshold region is a convincing technique to achieve low power in digital circuits. The problem is that process variability severely impacts the performance of circuits operating in the sub-threshold domain. In this paper, we evaluate the sub-threshold sizing methodology of [1,2] on 40 nm and 90 nm standard cell libraries. The concept of the proposed sizing methodology consists of balancing the mean of the sub-threshold current of the equivalent N and P networks. In this paper, the equivalent N and P networks are derived based on the best and worst case transition times. The slack available in the best-case timing arc is reduced by using smaller transistors on that path, while the timing of the worst-case timing arc is improved by using bigger transistors. The optimization is done such that the overall area remains constant with regard to the area before optimization. Two sizing styles are applied, one is based on both transistor width and length tuning, and the other one is based on width tuning only. Compared to super-threshold libraries, at 0.3 V, the proposed libraries achieve 49% and 89% average cell timing improvement and 55% and 31% power delay product improvement at 40 nm and 90 nm respectively. From ITC (International Test Conference 99) benchmark circuit synthesis results, at 0.3 V the proposed library achieves up to 52% timing improvement and 53% power savings in the 40 nm technology node.

Keywords: sub-threshold; process variation; library characterization; standard cell; sizing methodology; low power

---

1. Introduction

Low voltage digital design, especially near/sub-threshold design, is becoming more popular in application domains where performance is not the primary concern. More and more systems with low performance requirements are operated from a near/sub-threshold supply voltage in order to save power [3–7]. However, due to the fact that the gate voltage drive of the transistors operating in the sub-threshold domain is small, standard logic cells become more sensitive to process variations. Commercial cell libraries are designed and characterized for super-threshold voltage operation. Without any optimization, most cells of such conventional libraries will not have a robust operation in the presence of process variability at a low operating voltage. Therefore, careful sizing of standard cells working at low voltage is needed. In [1], the optimization procedures to size standard cells are explained. In [2], the standard cell libraries optimized for sub-threshold operation are presented. This paper extends the work of [1,2]. Here, the sizing methodology and sizing methods are explained using a CMOS 40 nm low power process as an example. Benchmarking of the libraries is carried out using both a CMOS 90 nm and a CMOS 40 nm low power process. ITC benchmark circuit synthesis results are presented as well.

Unlike conventional "super-threshold" cell sizing methods [8,9], the proposed balancing-based sizing method focuses on the statistical distribution of the drain-source current, rather than the current itself. In the proposed approach, the variation of the current is taken into consideration when sizing the standard cells by balancing the mean current of the equivalent N and P networks. The way of finding the equivalent N and P networks is based on timing arcs. The transition paths within the standard cells are different for distinct input patterns. The longest path, which has the worst delay, is defined as the worst-case transition path; the shortest path, which has the best delay, is defined as the best-case transition path. The transistors of the worst-case and the best-case transition paths are balanced in two possible ways: (i) transistor width and length tuning; and (ii) transistor width tuning only. In one case both the channel length and width of the transistor are optimized to have a better performance at low voltages, since in the sub-threshold regime, increasing the channel length has a positive impact on timing and timing variation [8]. Therefore, by increasing the transistor's length and by tuning the width [10] we are able to size the cells in the sub-threshold regime with two degrees of freedom. The second optimization approach, width tuning only, targets better timing and variation from the sub-threshold to the super-threshold regions.

Taking into account transistor sizing effects in sub-threshold [8], the balancing-based cell sizing methodology is presented in Section 2. Moreover, Section 2 also explains the standard cell optimization methods and how they can be applied to complex cells. A 163 standard cells library was designed and characterized using the proposed sizing methods in two technology nodes; the results are shown in Section 3. The evaluation of these libraries is presented in Section 4. Furthermore, to benchmark the libraries in the 40 nm technology node, ITC benchmark circuits are used to test the performance and variability of different libraries. The results are shown in Section 5. Section 6 concludes the paper.

2. Sub-Threshold Cell Sizing Methodology

Several relevant research results have been presented about sub-threshold sizing. In [3,4], the authors calculate the optimum supply voltage to minimize energy consumption. It is also claimed that, theoretically, minimum sized cells are optimal for energy reduction. In this paper it is shown that under speed constraints, and when process variability is taken into account, this is not the case. In [11], the authors explain the benefit of technology choices, power supply scaling, and body bias adaptability for circuits working in the sub-threshold regime. It is implied that standard cell timing could be improved using the mentioned design techniques. The concept of sub-threshold logical effort for complex gate sizing is presented in [9]. Particularly interesting is a closed form current equation derived for stacked transistors in relation to other transistors in the same stack. Compared to [3,4,9], our sizing approach focuses on narrowing the current/delay distribution spread and on increasing the performance through a new balancing theory that slows down fast transistors and *vice versa*. In [8], the transistor reverse short channel effect (RSCE) is used for device sizing optimization, where the channel length is increased to have an optimal threshold voltage which makes the transistors have a higher current, be less sensitive to random variations, and to have a smaller area. With a higher current and a lower gate capacitance, the delay and power are both reduced. Furthermore, in [8], the channel lengths of the NMOS and PMOS are increased to achieve the maximum currents for both NMOS and PMOS transistors. Unlike [8], our sizing optimization does not always lead to the maximum active current for both the NMOS and PMOS transistors. Only the transistors on slower timing arcs are allowed to be upsized, the ones on faster timing arcs are down sized to save area. In [12], a standard cell library in 65 nm is presented, where by upsizing the channel length of all transistors in a given cell, the energy per operation value is reduced by about 15%. In this paper, the standard cells are tuned individually, with various length and width selections to have balanced transition currents. Reference [13] presents a searching algorithm based on multiple objectives through a free space search to optimize one cell. The approach is exhaustive and suitable for single cells, but the searching effort is very large for a complete library. Unlike [11], our optimization targets balancing the mean P and N currents and takes into account the impact of process spread. In [14], a 45 nm standard cell library optimized for 0.35 V is proposed. The proposed PMOS-to-NMOS transistor ratio optimization is based on the optimal energy-delay product, not on balanced rise and fall times. In our work, the rise and fall times are balanced taking into account the effect of process variations.

Overall, in this section, a new statistical formulation [1] to size standard cells is introduced. The differences of the proposed work from other sizing methods are that in our work, the threshold voltage variation is treated as one of the statistical parameters in the current/delay equation, and the cells are optimized to have balanced current/delay distributions. The proposed sizing approach is derived from the observation that the transistor's current distribution in the sub-threshold regime follows a Log-Normal spreading, whereas conventional sizing treats the transistor's current as a Normal distribution. Considering the above-mentioned fact and the observation that process variability can be mapped onto threshold voltage variability with a first order approximation, a balancing based sizing methodology is developed for robust standard cell design.

2.1. Sub-Threshold Current Distribution Model

The sub-threshold region is often called the weak inversion region [15], partly because in the sub-threshold region, the transistor is neither completely turned on nor turned off. In digital circuits, the sub-threshold current is the parasitic leakage, ideally zero. By reducing the voltage supply to sub-threshold, and by letting the transistor operate in weak inversion, the power consumption can be reduced quadratically [16]. Transistors operating in the sub-threshold regime obey an exponential dependence on the gate drive voltage [8]:

$$I = \mu C \frac{W}{L} e^{1.8} U^2 e^{\frac{V_{gs}-V_{th}}{nU}} (1 - e^{\frac{-V_{ds}}{U}}) \tag{1}$$

where μ is the mobility; C is the oxide capacitance; n the sub-threshold slope factor; and U is the thermal voltage. V_{gs} is the gate to source voltage; V_{ds} is the drain to source voltage; V_{th} is the threshold voltage, consists of zero biasing voltage, terminal voltages and device size effects [17]. From Equation (1), one can see that the current has an exponential relationship with the gate-to-source voltage and the threshold voltage of the transistor.

In sub-threshold, the probability distribution function (PDF) of the current obeys a Log Normal distribution. If the supply voltage is reduced to the sub-threshold level, the widely distributed current will lead to a wide transistor delay spread. Therefore, an optimization based on a super-threshold current distribution will not guarantee a robust behavior in the sub-threshold regime. We consider the V_{th} as a Normal distribution and model the distribution of the transistor current using [18,19] as follows:

$$E[I] = \mu C \frac{W}{L} e^{1.8} U^2 e^{\frac{V_{gs}-E[V_{th}]}{nU} + \frac{Std^2[V_{th}]}{2(nU)^2}} (1 - e^{\frac{-V_{ds}}{U}})$$

$$Std^2[I] = (e^{\frac{Std^2[V_{th}]}{(nU)^2}} - 1)(E[I])^2 \tag{2}$$

where $E[]$ stands for the mean value and $Std[]$ stands for the standard deviation. In this model $E[V_{th}]$ and $Std[V_{th}]$ are regarded as technology parameters for a given W and L set. With the width and length tuning, $E[V_{th}]$ and $Std[V_{th}]$ also change accordingly due to RSCE. Therefore, depending on the range of W and L, different distributions of the V_{th} are used in the sizing model.

2.2. Sub-Threshold Cell Balancing Method

In traditional CMOS design, the transistor geometry ratio (W/L) of the pull-up PMOS network to the pull-down NMOS network is carefully tuned to compensate for the difference between the mobility of electrons and holes. This ratio is derived from balancing the rise/fall-time delays and minimizing the propagation delay.

In sub-threshold, it is more about equalizing the strength of the pull-up and the pull-down network that directly affects the functional correctness and the minimum V_{DD}. In the proposed sizing methodology, the ratio of the pull-up to pull-down transistors is determined by the balance between the current distributions of the PMOS and NMOS transistors. The difference with regard to the conventional sizing approach is that the current spread caused by the V_{th} variation is taken into account.

The proposed sizing methodology includes a transition-based approach in which the worst rise and fall times are improved by compromising the best rise and fall times. In this way, there is more room to improve the worst-case performance of the cells without area penalty.

Basically, the mean currents of the PMOS and NMOS networks are made equal, *i.e.*, $E[I_n] = E[I_p]$. From this, one can derive [1]:

$$\frac{W_n L_p}{W_p L_n} = \alpha e^{\frac{E[V_{thn}]-E[V_{thp}]}{nU}} e^{\frac{Std^2[V_{thp}]-Std^2[V_{thn}]}{2(nU)^2}} \tag{3}$$

where $\alpha = \mu_p C_p / \mu_n C_n$ is a technology parameter defined by the mobility and oxide capacitance of the NMOS and PMOS transistors. α is also used as the conventional sizing factor. Given the V_{th} mean and variance values, Equation (3) serves as the current balancing equation. The NMOS and PMOS current distributions can be closely matched based on Equation (3).

Figure 1 displays results of Monte Carlo simulations (CMOS 40 nm, 0.3 V power supply) of the normalized active current distributions of the NMOS and PMOS transistors of an inverter of strength 2 (INVD2). In the remaining of the paper the same commercial CMOS 40 nm technology is used as a reference. The current distributions of the NMOS and PMOS transistors can be closely matched, following Equation (3). Before balancing, the widths of the NMOS/PMOS are 0.62 μm/0.82 μm with fixed length of 0.041 μm. After balancing, the widths are 0.31 μm/0.60 μm and the lengths are 0.1 μm/0.044 μm, respectively. Note that the current distribution of the PMOS transistor is improved whereas the current of the NMOS transistor is weakened. In this case, the worst-case current distribution of the INVD2 is improved by reducing the best-case current. After the current balancing, the area of the INVD2 stays the same as before the balancing method is applied.

Figure 1. Normalized transistor current distributions in CMOS 40 nm. (**a**) Current distribution before balancing; (**b**) current distribution after balancing.

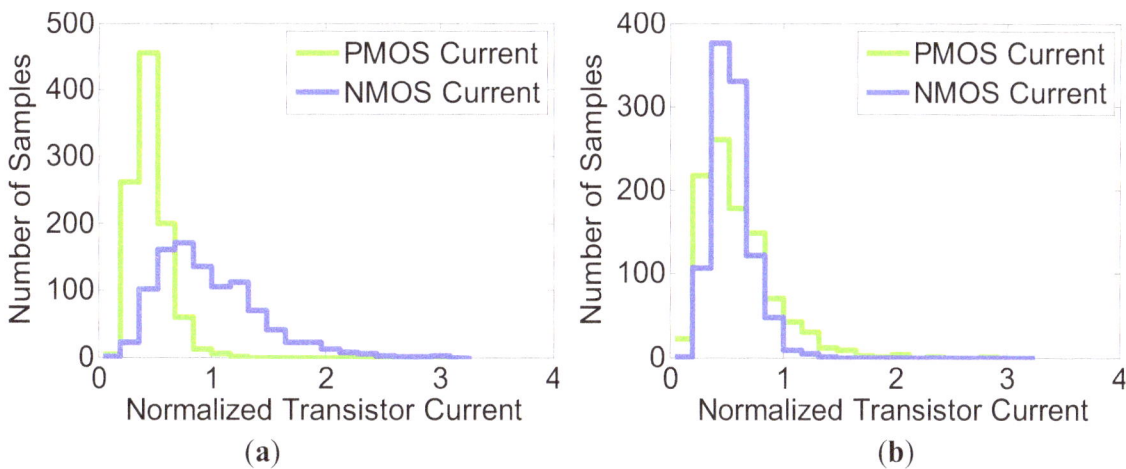

This balancing equation allows us to balance the rise and fall current distribution of the inverters without area penalty.

2.3. Stack Sizing Model

The magnitude of the current flowing through a transistor stack depends on the number of transistors and the size of each transistor. Without loss of generality, consider a transistor stack as depicted in Figure 2.

Figure 2. PMOS stack schematic.

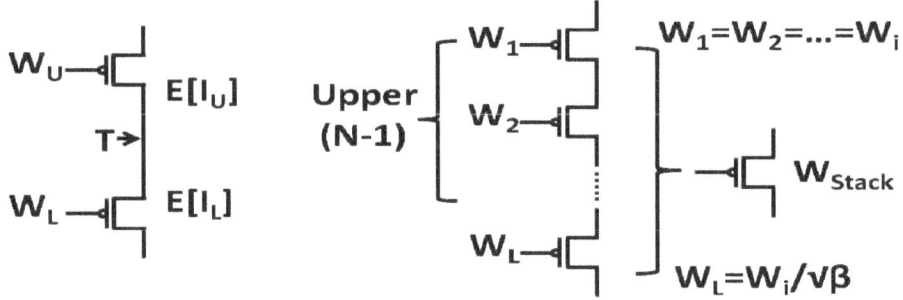

Let us enumerate this stack of PMOS transistors in descending order as a function of their proximity to the power supply V_{DD}. Similarly, consider a stack of NMOS transistors enumerated as a function of their proximity to Ground. Simulation results show that the upper $(N-1)$ PMOS transistors [lower $(N-1)$ NMOS transistors] have a similar impact on the current behavior of the stack. Therefore, let these $(N-1)$ transistors have equal sizes. Using the results of [9,20] to calculate the equivalent transistor width of the stack, W_{Stack}, the mean current of N transistors in a stack is calculated as follows [1]

$$E[I_{stack}] = K_E W_{Stack} e^{\frac{V_{dd}-E[V_{th}]}{nU}+\frac{Std^2[V_{th}]}{2(nU)^2}}$$
$$W_{Stack} = \frac{\beta W_L}{\left(1+\beta W_L\left(\sum_1^{N-1}\frac{1}{W_i}\right)\right)}; \beta = e^{\frac{-\lambda V_{dd}}{nU}} \tag{4}$$

where K_E is a technology fitting parameter and λ is the DIBL effect coefficient [9]. To simplify the calculation of the equivalent transistor size of the stack, the length of each transistor in the stack is held fixed. Let the width of all $(N-1)$ transistors be W_i and the width of the remaining transistor be $W_i/\sqrt{\beta}$, as shown in Figure 2. The width of the equivalent transistor is denoted as to W_{Stack}. The same procedure holds for NMOS transistors.

The variance of the stack is determined by the variance of each transistor in the stack. Since each transistor has the same impact on the total variance, the stack variance is the sum of the variances of each transistor divided by the square of the number of transistors in the stack [18].

$$Std^2[I_{stack}] = \frac{1}{N^2}\left(Std^2[I_L] + \sum_{i=1}^{N-1} Std^2[I_i]\right)$$
$$Std^2[I_{stack}] = \left(\frac{\beta(\sum_1^{N-1}W_i)+W_L}{K_{Std}N^2 W_{stack}}\right)(e^{\frac{Std^2[V_{th}]}{(nU)^2}} - 1)E^2[I_{stack}] \tag{5}$$

where K_{Std} is also a technology dependent fitting parameter. With Equations (4) and (5), one can easily derive the optimal stack width ratio for the stack's maximum current or minimum current spread. To achieve the maximum current, the lower PMOS (upper NMOS) transistor needs to be sized

$1/\sqrt{\beta}$ times smaller with regard to the upper PMOS (lower NMOS) transistors. The variation of the current stack can be written as:

$$\frac{Std[I_{stack}]}{E[I_{Stack}]} \propto \sqrt{\frac{\left(\beta\sqrt{\beta}(N-1)+1\right)\left(1+\sqrt{\beta}(N-1)\right)}{K_{Std}N^2\beta}} \tag{6}$$

Equation (6) helps to understand how many transistors can be stacked for given current variation and area constraints. Ultimately, this is a very important criterion for robust operation. To quantify this observation, 3000 Monte-Carlo simulations were run for 2, 3, 4, and 5 NMOS transistors in a stack working at 0.3 V and at room temperature (unless mentioned all the Monte-Carlo simulations are at 0.3 V and at room temperature). The results are shown in Table 1. The length of each transistor is held fixed to 0.04 μm, and the total width for each simulation set-up is set to 3 μm to keep the area constant. In Table 1 it is shown that Equation (6) predicts correctly the trend of the variation. The mismatch between the calculation and the simulation values is because V_{th} variation is treated as a given technology dependent parameter for given sizing (source bulk modulation is not taken into account). Table 1 is also an indicator of the large current variability when many transistors stacked transistors are used in the sub-threshold regime.

Table 1. Current variation in series-connected transistors @ 40 nm.

Number of transistors in series	Simulation results		Normalized $Std[I]/E[I]$	Calculation from Equation (6)
	$E[I]$ (A)	$Std[I]/E[I]$		
2×0.50 μm	2.31×10^{-8}	42.35%	1	1
3×0.33 μm	1.39×10^{-8}	53.03%	1.252	1.237
4×0.25 μm	1.11×10^{-8}	58.68%	1.386	1.401
5×0.20 μm	0.95×10^{-8}	66.18%	1.563	1.573

2.4. Parallel Sizing Model

The resulting PDF current of N parallel-connected transistors is the sum of their Log-Normal current distributions. The sum of Log-Normal distributions with the same variance can be approximated by one Log-Normal distribution [21]. A correlation factor ρ_p for V_{th} needs to be introduced to improve the accuracy of the model. This correlation factor was not needed in series-connected transistors because in that case the source-bulk modulation overshadows the correlation. The mean and variance of the current of N identical parallel connected transistors is [1]

$$E\left[I_{para}\right] = NK_{Ep}\frac{W_{one}}{L}e^{\frac{V_{gs}-E[V_{th}]}{nU}+\frac{Std^2[V_{th}]}{2(nU)^2}+\frac{N^2}{\rho_p}}$$

$$Std^2\left[I_{para}\right] = (e^{\frac{Std^2[V_{th}]}{(nU)^2}+\frac{2N}{\rho_p}} - 1)\left(E[I_{para}]\right)^2/N^2 \tag{7}$$

where W_{one} is the width of one single transistor, $K_{Ep} = \mu C e^{1.8}U^2(1 - e^{\frac{-V_{ds}}{U}})$ and $\rho_p \propto Std^2[V_{th}]$. The equivalent width for parallel transistors can be calculated from Equation (7) [1].

$$W_{Para} = \gamma(N)W_{one}$$

$$\gamma(N) = Ne^{\frac{N^2}{\rho_p}} \tag{8}$$

Hence the width of a single transistor, which has the same mean current as the one of N transistors in parallel, is $\gamma(N)$ times the width of the transistors in parallel.

To quantify our model, 3000 Monte-Carlo simulations were run for 1 to 6 NMOS transistors connected in parallel, with a total width of 1.20 μm. The simulation and calculation results are shown in Table 2. It is worth observing these results in more detail [22]. Namely, the joint correlated Log-Normal distribution indicates that the mean current is bigger than that of the uncorrelated sum of individual transistor currents [18,21]. This implies that for the sub-threshold regime it could be advantageous to layout parallel-connected transistors as the current drive is higher.

Table 2. Mean current of parallel-connected transistors in CMOS 40 nm.

Number of parallel transistors	Simulated $E[I]$ (A)	Normalized $E[I]$	Calculation from Equation (7)
1×1.20 μm	1.18×10^{-7}	1.00	1.00
2×0.60 μm	1.33×10^{-7}	1.13	1.12
3×0.40 μm	1.41×10^{-7}	1.19	1.24
4×0.30 μm	1.52×10^{-7}	1.29	1.36
5×0.24 μm	1.71×10^{-7}	1.45	1.48
6×0.20 μm	1.91×10^{-7}	1.62	1.61

2.5. Complex Cell Translation

Complex cells can be sized by finding equivalent transistor sizes from reducing stack and parallel arrangements to their equivalent reference transistors. Note that when the stack or parallel arrangement is reduced to the equivalent reference transistors, the distribution parameters, the mean and standard deviation of the arrangements are also calculated by the equations shown above.

Without loss of generality, a complex cell as the one depicted in the left part of Figure 3 is used to explain how the cell is "reduced". The equivalent sizes of the transistors in series or parallel connection can be determined by two rules as depicted in

Algorithm 1.

1. **If** *n transistors in Parallel*
2. **Then** *Size* of parallel transistors:
3. $W_1 = W_2 = \cdots = W_n$
4.   Parallel Equivalent Size: *Equation* (8)
5. **If** *m transistors in Series*
6. **Then** *Size* of transistors in stack:
7. $W_{L1} = W_{L2} = \cdots W_{L(m-1)} = \sqrt{\beta} W_U$
8.   Stack Equivalent Size: *Equation* (4)
9. *U means next to output node; L means away from output node.

Figure 3. Complex cell translation example.

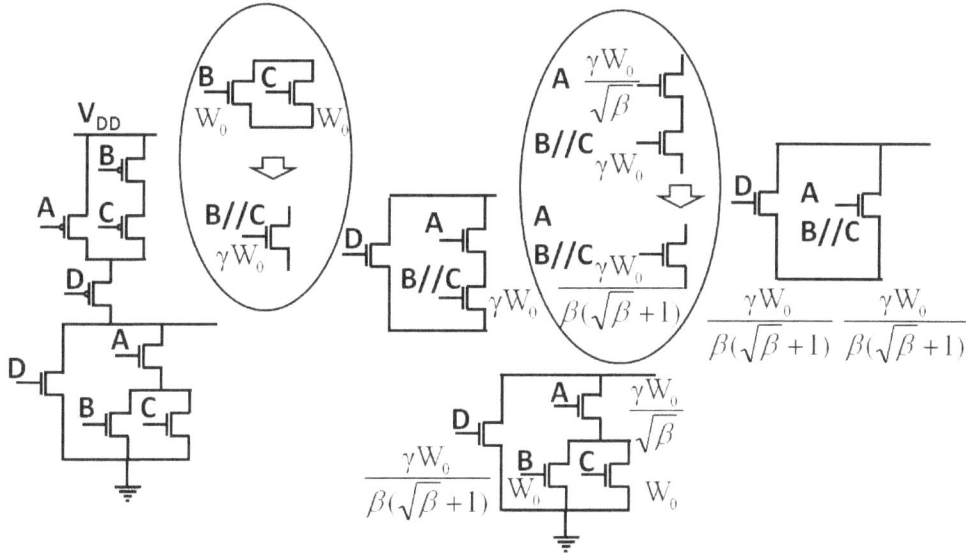

In the right part of Figure 3, the NMOS network is used as an example to show how the sizing ratio is determined by two If-Then rules. The translation starts with the parallel-connected NMOS transistor B and C. The initial sizes of size of B and C are equal to W_0 as the unit size of the N network. Then, the size of the equivalent transistor of B // C is γW_0 according to Equation (8). With transistor A in series connection, the size of A can be defined by the second if then rule, as $\gamma W_0 / \sqrt{\beta}$. The equivalent size of A, B // C is defined by Equation (4) as $\gamma W_0 / \beta (\sqrt{\beta} + 1)$. The size of transistor D is equal to the size of the equivalent parallel-connected transistors. A similar procedure can be followed to size the transistors of the P network.

The sizing approach that we just outlined can guarantee the maximum PDF current within the N/P networks. The immediate follow up step is to balance the fall and rise delays across the N/P networks according to Equation (3). For the transistors shown in Figure 3, the balancing Equation (3) is applied between the worst-timing transition path in the N network (transistors A and C) and the best-timing transition path in the P network (transistors A and D), and between the best-timing transition path in the N network (transistor D) and the worst-timing transition path in the P network (transistors B, C, and D). The equivalent transistor of the transistors on the best/worst timing transition path within N/P network is determined by Equations (4) and (8). For example, consider the worst-timing transition path in the N network consisting of transistors A and C. Following Equation (4), we substitute W_0 for W_i and $\gamma W_0 / \sqrt{\beta}$ for W_L. Then, the equivalent size of the worst-timing transition path in the N network becomes $\sqrt{\beta} \gamma W_0 / (\sqrt{\beta} + 1)$. This is balanced against the equivalent transistor resulting from best timing transition path in the P network using Equation (3) to find the actual width values of W_0 for the N and P networks. Other combinations of equivalent transistors on the best/worst timing transition path of the N/P network can be derived accordingly.

In this paper the libraries are targeted at balancing the worst-case rise and fall transitions.

When both width tuning and channel length tuning are considered, the library performs well at near threshold supply voltages [13]. For higher supply voltages, the benefit of having non-minimum channel length decreases. On the other hand, the library in which only width optimization is used has a constant improvement over a wide voltage ranges as compared to the reference library. Therefore, the former should be used for digital blocks mainly operating in sub-threshold region, while the latter should be used for blocks, which are working in a wide voltage range from sub-threshold voltage to nominal supply voltage. The cell area constraint is set to be the same for both libraries and equal to the corresponding "super-threshold" cell area. The differences are only on individual transistor sizes, so there is no extra area cost.

3. Library Characterization

To benchmark the sizing methods, all libraries (two at 40 nm technology and three at 90 nm technology) are characterized for worst-case timing and power from 0.3 V to 1.2 V in 0.1 V steps based on the layout extracted standard cell netlists (including parasitic). As the super-threshold cells will not function properly under 0.3 V, 0.3 V is set as the lowest characterization voltage to have a fair comparison with the proposed libraries. The characterization is done in SS process corner at room temperature with slew and loading ranges appropriate to the supply voltage. Since the area is constrained to be the same as the corresponding super-threshold libraries, the loading stays the same as the one in the super-threshold libraries. To define the slew range, a single drive strength inverter with loading specified by the commercial super-threshold libraries is simulated. The appropriate slew for each voltage is determined by matching rise/fall times of the input node and fall/rise times of the output node respectively.

Both ELC and Altos of Cadence are used for library re-characterization. Altos is used for the 40 nm library, and ELC is used for the 90 nm one. The simulation engine is Spectre. The results of the library characterization are stored according to the commercial liberty format [23]. The timing information of each pin of each cell is presented in four matrices: rise time represents the rising slew, rise transition represents the transition time when the output rises. Similarly, fall time and fall transition represent the delay when the output falls. The characterized libraries follow a 7×7 timing and power template. Each matrix consists of 49 values for seven different slew times and seven loading parameters. A similar format also applies to power information.

4. Library Comparisons

Since the values of slew and loading parameters differ over two orders of magnitude in these matrices, it is not convenient to carry out a straightforward comparison. Instead, the average value of each matrix is used to represent the delay and power, called as pin-delay and pin-power parametric.

The pin-delay and pin-power values are used to compare the proposed sub-threshold libraries to the "super-threshold" library at different voltages. The comparisons are carried out on a CMOS 90 nm and on a CMOS 40 nm SVT technology. In Figure 4, the voltage scalability of different libraries at 90 nm is presented. Timing improvement is calculated by comparing the delay value of each cell in sub-threshold to the corresponding cell in the super-threshold library, and the average of all improvements are compared. The library with width and length tuning shows around 49% better timing at 0.3 V, and when the voltage increases to 0.65 V, the improvement drops to 0. Above 0.65 V the library with width and length tuning works slower than the "super-threshold" library. The library with width tuning only shows 10% to 11% better average timing from 0.3 V to 1.2 V compared to the "super-threshold" library.

Figure 4. Average cell timing improvement of different voltages in CMOS 90 nm.

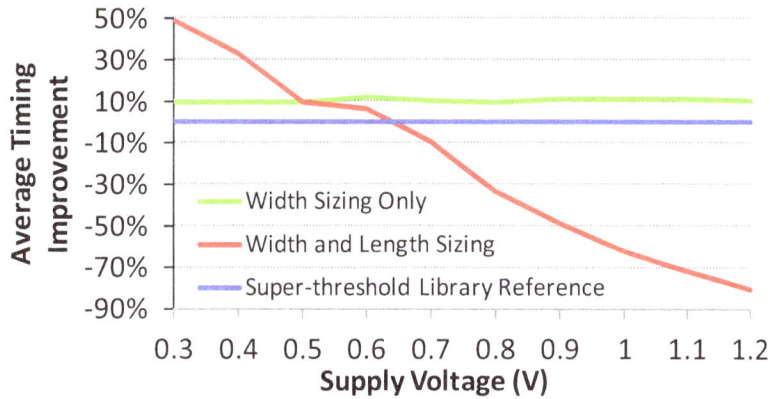

In Figure 5, the width and length tuning library is compared to the "super-threshold" library at 0.3 V. The max cell delay is the maximum value of the pin-delay of each cell. It actually shows the worst average transition of each cell. The corresponding pin-delay and pin-power are used to compare the power delay product (PDP) of each cell. The max cell and the max cell PDP are compared in each technology node.

Figure 5. Normalized max cell delay and PDP comparison in CMOS 90 nm and 40 nm. (**a**) Delay comparison at 90nm; (**b**) PDP comparison at 90nm; (**c**) delay comparison at 40nm; (**d**) PDP comparison at 40nm

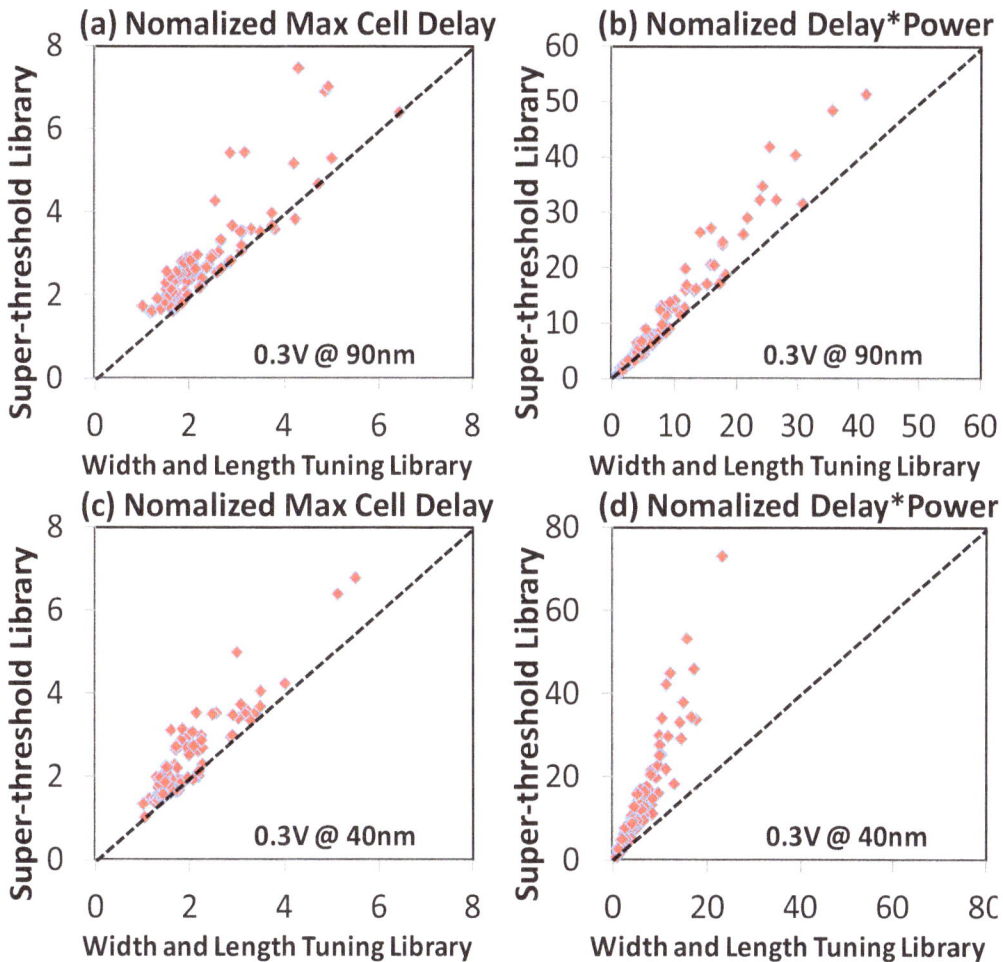

One can see that most of the cells from the width and length tuning library lie above the reference 45 degree dashed line, which means that the cells from the width and length tuning library have better timing properties. Those cells that lie on the reference line are the minimum sized cells, which cannot be further optimized using the proposed balancing-based sizing method. Following the proposed sizing method, the complex cells and cells with larger drive strength have better performance compared to the rest of the logic cells.

On average, the 90 nm cells with width and length tuning have 38% better timing for worst case transitions without introducing extra area cost. On average the cells from the width and length tuning library achieve 31% better PDP at worst transition. In the 40 nm technology node, the width and length tuning library cells have 49% average timing improvement for worst-case transitions and 55% better average PDP compare to the super-threshold library reference at 40 nm.

Three thousand Monte Carlo simulations have been done for each cell to compare their timing variation at 0.3 V. The results of the delay, variation and area of the cells are shown in Figure 6. The marker size shows the area of the cell. As known, bigger cells have less variation [24]. However, in the figure, this is not always true for all the cells; most of the cells lie in the standard deviation/mean range from 50% to 70%. There is no clear indication that, increasing the area will lead to variation savings in the sub-threshold region.

In Figure 6, we see that our cells are mainly distributed in the lower left corner, which means that the performance and the robustness of our cells are better than the cells of the super-threshold library, as expected. On average, the cells that follow the width and length tuning method have 11% variation savings and 2.17× performance improvement at 40 nm. Among all the cells compared, the width and length tuning have maximally 45% variation savings for a two input NOR gate NR2D1 and 4.12× maximum performance improvement for the NR2XD8 without any area penalty.

Figure 6. CMOS 40 nm libraries cell delay variation and area comparison at 0.3 V. The values in the figure are normalized to the minimum mean delay of each super-threshold library.

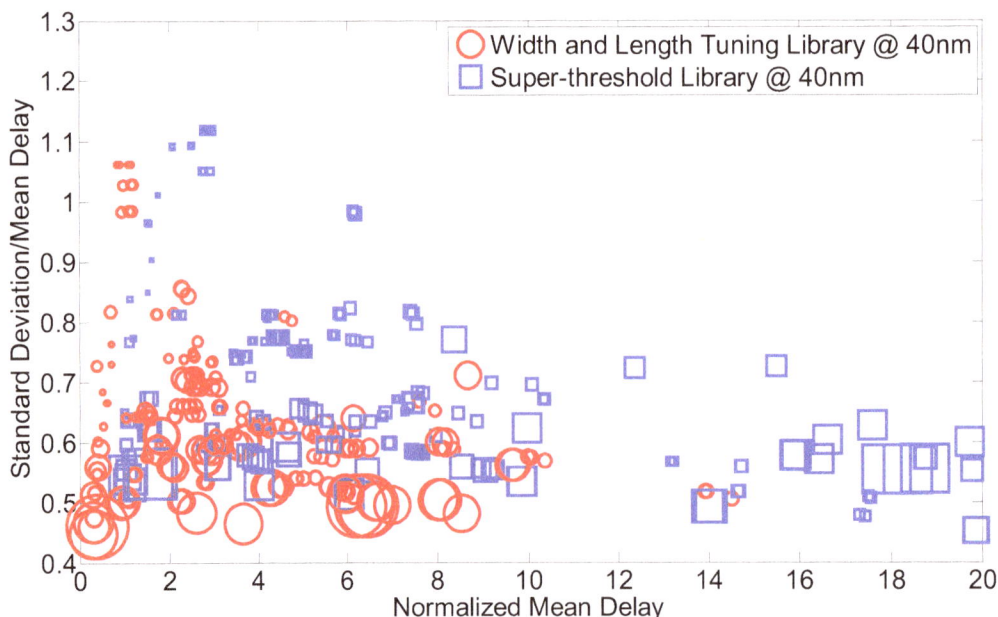

5. Circuit Synthesis Comparisons

5.1. ITC B14 Benchmark

We look here first in detail at synthesis results of the B14 circuit from ITC benchmark circuit [25].

We extracted the critical paths generated by each different library at 0.3 V, and then applied 1000 Monte Carlo simulations are used to generate the delay distributions of each critical path to compare the variability of different libraries. The results are shown in Figure 7.

Figure 7. Critical path delay distribution comparisons in CMOS 40 nm.

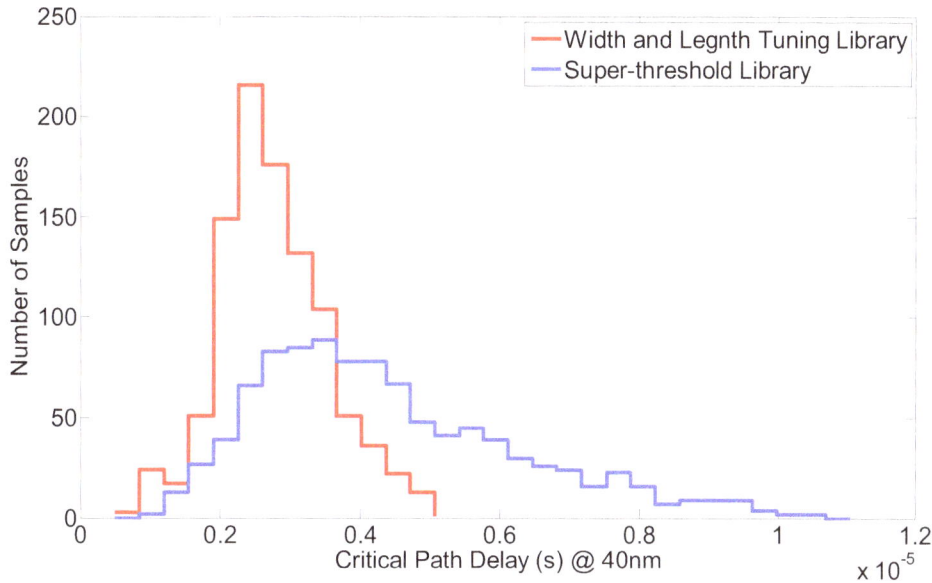

As can be seen from Figure 7, the critical path delay follows a Log-Normal distribution. Without any sizing optimization, the critical path has a wide distribution with a long tail as the blue line shows. One can see that the delay distribution of the critical path of the proposed library cells is left shifted and narrowed down, where the mean delay decreases from 4.25 μs to 2.59 μs, and the variation is reduced from 44% to 30%.

In Figure 8, the synthesized delay *versus* area trend at 0.3 V is compared. The three black arrows show different constraints. With the width and length-tuning library, the circuit can work at faster speed. Arrow C indicates that, when delay is a constrain, the circuit synthesized by the width and length tuning library requires 14% less area as compared to the circuit synthesized with the super-threshold library. When area is the constraint (arrow B), the circuit synthesized by the width and length-tuning library is 1.8× faster. Without any constraints, the circuit can be sped up 2.1× with 1.08× area compared to the circuit synthesized by the super-threshold library, as indicated by arrow A.

Figure 8. Synthesized circuit delay and area comparison in CMOS 40 nm at 0.3 V.

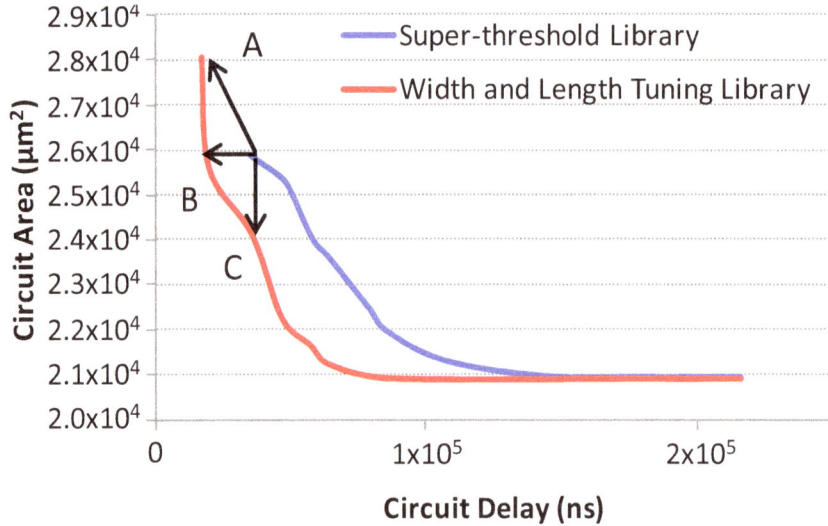

5.2. ITC Benchmark Circuits

ITC benchmark circuits [25] were synthesized for minimum delay to compare 40 nm libraries at 0.3 V. The delay, area, and power information are shown in Tables 3–5. In Table 3, the speed of the circuit synthesized by the proposed library is pushed to the highest possible value like the arrow A in Figure 8. Table 4 shows the delay improvement and power savings when the area is constrained as arrow B in Figure 8. Table 5 shows the area and power saving when the same target delay is applied as the arrow C in Figure 8.

Table 3. ITC benchmark circuit synthesis results.

	Delay (ns)			Area (μm^2)			Total Power (nW)		
	Super-threshold library	Width and length tuning	%	Super-threshold library	Width and length tuning	%	Super-threshold library	Width and length tuning	%
B01	850	480	43.5	320	334	−4.4	0.502	0.308	38.6
B02	780	450	42.3	213	227	−6.6	0.237	0.161	32.1
B03	880	510	42.0	582	660	−13.4	0.229	0.164	28.4
B04	1170	630	46.2	2120	2525	−19.1	1.267	0.865	31.7
B05	1820	1030	43.4	3118	3664	−17.5	1.336	0.920	31.1
B14	3600	1720	52.2	25866	28056	−8.5	3.795	2.780	26.7

ITC benchmark results show that, in the 40 nm technology node, the circuits synthesized by the proposed width and length-tuning library have better timing, less area, and less power consumption when compared to the super-threshold library at 0.3 V. For the delay driven comparison shown in Table 3, we observe a maximum timing improvement of 52% and power savings of 39%. If the same area constraint is applied, the maximum timing improvement is 44% and the power saving is 41%. When the delay target is set the same for both libraries, the width and length tuning library achieves up to 24% area savings, and 53% power savings.

Table 4. ITC benchmark circuit synthesis results with the equal area constraint.

	Delay (ns)			Area (μm²)			Total Power (nW)		
	Super-threshold library	Width and length tuning	%	Super-threshold library	Width and length tuning	%	Super-threshold library	Width and length tuning	%
B01	850	500	41.2	320	315		0.502	0.298	40.6
B02	780	490	37.2	213	204		0.237	0.148	37.6
B03	880	750	14.8	582	555		0.229	0.138	39.7
B04	1170	810	30.8	2120	2077		1.267	0.765	39.6
B05	1820	1200	34.1	3118	3114		1.336	0.826	38.2
B14	3600	2000	44.4	25866	25614		3.795	2.466	35.0

Table 5. ITC benchmark circuit synthesis results with same delay constraint.

	Delay (ns)		Area (μm²)			Total Power (nW)		
	Super-threshold library	Width and length tuning	Super-threshold library	Width and length tuning	%	Super-threshold library	Width and length tuning	%
B01	850	850	320	243	24.1	0.502	0.238	52.6
B02	780	780	213	177	16.9	0.237	0.144	39.2
B03	880	880	582	536	7.9	0.229	0.139	39.3
B04	1170	1170	2120	1671	21.2	1.267	0.877	30.8
B05	1820	1820	3118	2726	12.6	1.336	0.723	45.9
B14	3600	3600	25866	24121	6.7	3.795	2.852	24.8

6. Conclusions

In this paper, we presented an impact analysis of sub-threshold sized libraries against super-threshold sized ones. The proposed sizing methods were benchmarked against a library tuned for super-threshold operation in the 90 nm and 40 nm technology nodes. The simulation results of the ITC benchmark circuits show that the proposed width and length tuning library achieves up to 52% (average 45%) timing improvement and up to 38% (average 32%) power saving with 11% area overhead. When area is held constant, the maximum timing improvement figure drops to 44% (average 34%) and maximum power saving figure increases to 41% (average 38%). When timing is held constant, the maximum area saving is 24% (average 15%) and maximum power saving figure increases to 53% (average 39%).

References

1. Liu, B.; Ashouei, M.; Huisken, J.; de Gyvez, J.P. Standard Cell Sizing for Subthreshold Operation. In Proceedings of the 49th Design Automation Conference (DAC), San Fransico, CA, USA, 3–7 June 2012; pp. 962–967.

2. Liu, B.; de Gyvez, J.P.; Ashouei, M. Library Tuning for Subthreshold Operation. In Proceedings of the 2012 IEEE Subthreshold Microelectronics Conference (SubVT), Waltham, MA, USA, 9–10 October 2012; pp. 1–3.

3. Calhoun, B.H.; Wang, A.; Chandrakasan, A. Modeling and sizing for minimum energy operation in subthreshold circuits. *Solid-State Circ. IEEE J.* **2005**, *40*, 1778–1786.

4.   Kwong, J.; Ramadass, Y.; Verma, N.; Koesler, M.; Huber, K.; Moormann, H.; Chandrakasan, A. A 65nm Sub-Vt Microcontroller with Integrated SRAM and Switched-Capacitor DC-DC Converter. In Proceedings of the IEEE International Solid-State Circuits Conference (ISSCC), San Fransico, CA, USA, 3–7 February 2008; pp. 318–616.

5.   Seok, M.; Jeon, D.; Chakrabarti, C.; Blaauw, D.; Sylvester, D. A 0.27V 30MHz 17.7nJ/Transform 1024-pt Complex FFT Core with Super-Pipelining. In Proceedings of the 2011 IEEE International Solid-State Circuits Conference Digest of Technical Papers (ISSCC), San Fransico, CA, USA, 20–24 February 2011; pp. 342–344.

6.   Bol, D.; Kamel, D.; Flandre, D.; Legat, J.-D. Nanometer MOSFET Effects on the Minimum-Energy Point of 45nm Subthreshold Logic. In Proceedings of the 14th ACM/IEEE International Symposium on Low Power Electronics and Design, San Francisco, CA, USA, 19–21 August 2009; pp. 3–8.

7.   Kwong, J.; Chandrakasan, A.P. Variation-Driven Device Sizing for Minimum Energy Sub-Threshold Circuits. In Proceedings of the 2006 International Symposium on Low Power Electronics and Design (ISLPED), Tegernsee Germany, 4–6 October 2006; pp. 8–13.

8.   Kim, T.-H.; Hanyong, E.; Keane, J.; Kim, C. Utilizing Reverse Short Channel Effect for Optimal Subthreshold Circuit Design. In Proceedings of the 2006 International Symposium on Low Power Electronics and Design (ISLPED), Tegernsee, Germany, 4–6 October 2006; pp. 127–130.

9.   Keane, J.; Hanyong, E.; Tae-Hyoung, K.; Sapatnekar, S.; Kim, C. Subthreshold Logical Effort: A Systematic Framework for Optimal Subthreshold Device Sizing. In Proceedings of the 43rd Design Automation Conference, San Fransico, CA, USA, 24–24 June 2006; pp. 425–428.

10. Jun, Z.; Jayapal, S.; Busze, B.; Huang, L.; Stuyt, J. A 40 nm Inverse-Narrow-Width-Effect-Aware Sub-Threshold Standard Cell Library. In Proceedings of the 48th Design Automation Conference (DAC), San Diego, CA, USA, 5–9 June 2011; pp. 441–446.

11. Bol, D.; Flandre, D.; Legat, J.-D. Technology Flavor Selection and Adaptive Techniques for Timing-constrained 45nm Subthreshold Circuits. In Proceedings of the 14th International Symposium on Low Power Electronics and Design, San Francisco, CA, USA, 19–21 August, 2009; pp. 21–26.

12. Bol, D.; de Vos, J.; Hocquet, C.; Botman, F.; Durvaux, F.; Boyd, S.; Flandre, D.; Legat, J. SleepWalker: A 25-MHz 0.4-V Sub-mm^2 7-uW/MHzMicrocontroller in 65-nm LP/GP CMOS for low-carbon wireless sensor nodes. *Solid-State Circ. IEEE J.* **2013**, *48*, 20–32.

13. Blesken, M.; Lu, X; Tkemeier, S.; Ruckert, U. Multiobjective Optimization for Transistor Sizing Sub-threshold CMOS Logic Standard Cells. In Proceedings of 2010 IEEE International Symposium on Circuits and Systems, Paris, France, 30 May–2 June 2010; pp. 1480–1483.

14. Abouzeid, F.; Clerc, S.; Firmin, F.; Renaudin, M.; Sicard, G. A 45nm CMOS 0.35v-optimized Standard Cell Library for Ultra-low power Applications. In Proceedings of the 14th ACM/IEEE International Symposium on Low Power Electronics and Design, San Francisco, CA, USA, 19–21 August 2009; pp. 225–230.

15. Tsividis, Y. *Operation and Modeling of the Mos Transistor* (*The Oxford Series in Electrical and Computer Engineering*); Oxford University Press: New York, USA, 2004; pp. 62–96.

16. Wang, A.; Calhoun, B.H.; Chandrakasan, A.P. *Sub-Threshold Design for Ultra Low-Power Systems*; Springer: New York, USA, 2006; pp. 27–32.

17. *Avant, Star-Hspice User's Manual*; Synopsys: Mountain View, CA, USA, 2000; pp. 798–801

18. Crow, E.L.; Shimizu, K. *Lognormal Distributions: Theory and Applications*; Marcel Dekker: New York, NY, USA 1988; pp. 195–210.

19. Bo, Z.; Hanson, S.; Blaauw, D.; Sylvester, D. Analysis and Mitigation of Variability in Subthreshold Design. In Proceedings of the 2005 International Symposium on Low Power Electronics and Design, San Diego, CA, USA, 8–10 August 2005; pp. 20–25.

20. Al-Hertani, H.; Al-Khalili, D.; Rozon, C. A New Subthreshold Leakage Model for NMOS transistor Stacks. In Proceedings of the IEEE Northeast Workshop on Circuits and Systems, Montreal, Canada, 5–8 August 2007; pp. 972–975.

21. Fenton, L. The sum of log-normal probability distributions in scatter transmission systems. *Commun. Syst. IRE Trans.* **1960**; *8*, 57–67.

22. Gemmeke, T.; Ashouei, M. Variability Aware Cell Library Optimization for Reliable Sub-Threshold Operation. In Proceedings of the European Solid States Circuits Conference (ESSCIRC), Bordeaux, France, 17–21 September 2012; pp. 42–45.

23. Bhasker, J.; Chadha, R. *Static Timing Analysis for Nanometer Designs: A Practical Approach*; Springer: New York, USA, 2009; pp 26–43.

24. Pelgrom, M.J.M.; Duinmaijer, A.C.J.; Welbers, A.P.G. Matching properties of MOS transistors. *Solid-State Circ. IEEE J.* **1989**, *24*, 1433–1439.

25. Corno, F.; Reorda, M.S.; Squillero, G. RT-level ITC'99 benchmarks and first ATPG results. *Des. Test Comput. IEEE* **2000**, *17*, 44–53.

5

A Low Power CMOS Imaging System with Smart Image Capture and Adaptive Complexity 2D-DCT Calculation

Qing Gao and Orly Yadid-Pecht *

Department of Electrical and Computer Engineering, University of Calgary, AB T2N1N4, Canada; E-Mail: qgao@ucalgary.ca

* Author to whom correspondence should be addressed; E-Mail: orly.yadid.pecht@ucalgary.ca

Abstract: A novel low power CMOS imaging system with smart image capture and adaptive complexity 2D-Discrete Cosine Transform (DCT) is proposed. Compared with the existing imaging systems, it involves the smart image capture and image processing stages cooperating together and is very efficient. The type of each 8×8 block is determined during the image capture stage, and then input into the DCT block, along with the pixel values. The 2D-DCT calculation has adaptive computation complexity according to block types. Since the block type prediction has been moved to the front end, no extra time or calculation is needed during image processing or image capturing for prediction. The image sensor with block type decision circuit is implemented in TSMC 0.18 µm CMOS technology. The adaptive complexity 2D-DCT compression is implemented based on Cyclone EP1C20F400C8 device. The performance including the image quality of the reconstructed picture and the power consumption of the imaging system are compared to those of traditional CMOS imaging systems to show the benefit of the proposed low power algorithm. According to simulation, up to 46% of power consumption can be saved during 2D DCT calculation without extra loss of image quality for the reconstructed pictures compared with the conventional compression methods.

Keywords: CMOS imaging system; low power; smart image capture; adaptive complexity DCT

1. Introduction

Wide utilization of portable battery-operated devices in multimedia applications, such as cell phones, portable digital assistants (PDAs) and smart toys, has triggered a demand for ultra low-power image system. CMOS imaging technology has recently become a very attractive solution for these applications as they consume less power, and operate at higher speeds compared with CCD imaging technology [1].

Many low power designs for image capture were reported during the last decade [2–8]. A review of low power designs in CMOS image sensors at different levels is given in [2]. Some works aim at compensating the reduced signal to noise ratio and dynamic range caused by a low operating voltage [3]. In [4], SOI (Silicon-On-Insulator) technology is used instead of the traditional CMOS technology since it has smaller parasitic capacitance and reduced leakage current. In 2006, the first self powered image sensor was proposed by Fish *et al.* [5]. Then an optimized energy harvesting CMOS image sensor was proposed [6], where the photodetector itself can be used for power generation besides the PGPd. In [7,8] a block based dual VDD image sensor was proposed. It has dual supply voltages during image capture stage and the supply voltage is decided according to the block types.

After the image capture, energy-aware data compression is usually performed for efficient transmission. The Discrete Cosine Transform (DCT), and in particular the DCT-II, is often used in image/video processing such as JPEG still image compression, MJPEG, MPEG video compression due to its good energy compaction. However the DCT itself contains very computationally intensive matrix multiplications and therefore is power consuming. Numerous algorithms have been proposed attempting to minimize the number of additions and multiplications such as the Loeffler DCT [9–11] or even to replace multiplications with only add and shift operations, *i.e.*, Distributed Arithmetic (DA), Coordinate Rotation Digital Computer (CORDIC) and binDCT [12–14]. Also, data-dependent DCT algorithms have been introduced for low power purpose [15].

In all the existing CMOS imaging systems, the image capture stage and the compression stage are simply concatenated together. The DCT architectures without multiplier are usually time consuming and hardware-expensive. Some DCT designs subsample the area where pixels change less, but all the block type predictions are made during the digital image processing and require extra processing time and also extra memory space to store the image data during prediction. However, in our imaging system, these two stages are intertwined together. The block type prediction is moved to the front end, that is, the block type is estimated during image capture in an analog way (at the mean time during read out). It is more efficient in terms of power. In addition, no extra time is needed during image processing or image capturing for block type prediction. The 2D DCT calculation has adaptive data format and computation complexity depending on the block type.

The paper is organized as follows. In Section 2, the general algorithm and architecture of the proposed low power imaging system is described. Section 3 discusses the circuit implementation of the imager with the block decision circuit and the adaptive-complexity 2D-DCT. Performance regarding the image quality and the power consumption will be also given in Section 3. Finally, conclusions are presented in Section 4.

2. Proposed Low Power Imaging System with Smart Image Capture and Adaptive Complexity 2D-DCT

2.1. Traditional CMOS Imaging System with 2D-DCT Calculation

In a traditional CMOS imaging system, the image is first captured by the imager, and then the analog image data go through an Analog-to-Digital (ADC) conversion and then digital processing, such as image compression. The general diagram of a basic traditional digital camera system with 2D-DCT based compression is shown in Figure 1.

Figure 1. A traditional digital imaging system with 2D-Discrete Cosine Transform (DCT) calculation.

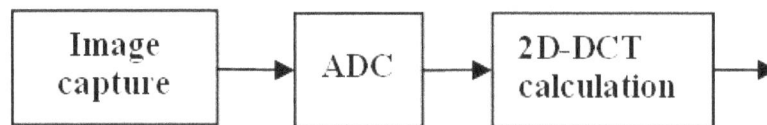

In hardware the N × N 2D-DCT can be realized by storing the output of the first 1-D DCT in a memory buffer line after line, then applying a second 1-D DCT transform on the columns of the results [16]. N is typically 8 in most of the applications, resulting in an 8 × 8 transform coefficient.

2.2. Proposed Low Power Imaging System with Smart Image Capture and Adaptive Complexity 2D-DCT Calculation

2.2.1. Architecture of the Proposed Low Power Imaging System

The architecture of the novel low power digital camera system with adaptive complexity 2D-DCT based calculation is shown in Figure 2. The system is mainly composed of an image sensor for smart image capture, an ADC for data format converting and an adaptive-complexity 2D-DCT calculation. In addition to capturing images as the traditional CMOS imagers, the image sensor in the proposed camera system contains a block type decision block that can compute the type for each block of 8 × 8 pixels according to the variance [17]. In order to simplify the implementation, we use the difference between the maximum and minimum values within a block as an approximation of the variance to represent how far a set of numbers is spread out, similar to what was done in [8]. A block with lower $Vmax - Vmin$ has a trend of lower variance. This is more accurate when the variance is small, so we use $Vmax - Vmin$ as an approximation of the variance to simplify the implementation. The complexity of the proposed 2D DCT calculations is dependent on the block type decided in the image sensor during image capture stage. Power saving is achieved with reduced computation during DCT for small variance blocks.

Figure 2. The proposed low power digital CMOS imaging system with smart image capture and adaptive complexity 2D-DCT calculation.

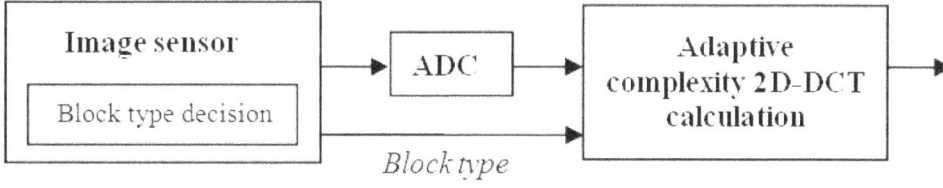

2.2.2. Adaptive Complexity Compression

As shown in Figure 3, for blocks with large *Vmax – Vmin* (object blocks), conventional DCT is performed. For the blocks with small *Vmax – Vmin* (background blocks), the differential data format is used instead of the pixel value itself during AC coefficients calculation. In addition, the resolution unit is considered as N × N to reduce the computation complexity. Here N can be selected from 1, 2, 4 and 8 according to different applications. In our implementation, N is chosen as 2 for now. With reduced spatial resolution, part of the computation can be skipped without loss of useful information. Because less bits are used during the AC coefficients computing and part of the calculation is skipped, power saving is achieved for background blocks.

Figure 3. Example of the proposed low power algorithm.

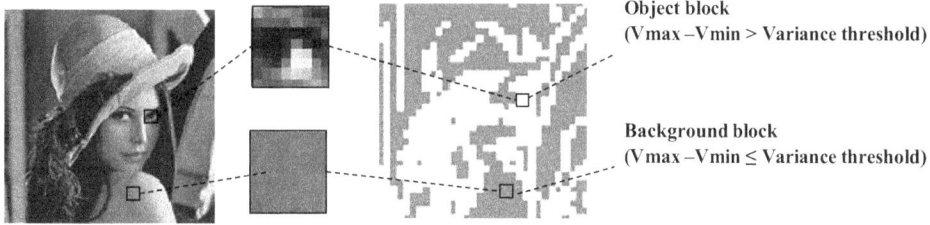

Object block
(Vmax −Vmin > Variance threshold)

Background block
(Vmax −Vmin ≤ Variance threshold)

Using vector form, the 8 × 8 DCT transform becomes $F = CXC^T$. Where X is the input matrix and F is the output results after DCT transform. C is the cosine coefficients matrix and C^T is the transpose coefficients matrix.

For background type blocks, the spatial resolution unit will be considered as 2 × 2 during calculation, therefore the row DCT transform becomes $F' = CX'C^T$, where C is the new cosine coefficients and C^T is the new transpose coefficients. X' is the new pixel block and F' is the corresponding result after DCT transform.

$$F' = CX'C^T = C_{8x8}X'_{8x8}C_{8x8}{}^T = C_{8x8}(B_{8x4}X''_{4x4}B_{8x4}{}^T)C_{8x8}{}^T$$
$$= (C'_{8x4})X''_{4x4}(C'_{8x4})^T \tag{1}$$

where $B_{8x4} = \begin{bmatrix} 1 & 1 & 0 & 0 & 0 & 0 & 0 & 0 \\ 0 & 0 & 1 & 1 & 0 & 0 & 0 & 0 \\ 0 & 0 & 0 & 0 & 1 & 1 & 0 & 0 \\ 0 & 0 & 0 & 0 & 0 & 0 & 1 & 1 \end{bmatrix}^T$, $X''_{4x4} = \begin{bmatrix} x0 & x2 & x4 & x6 \\ x16 & x18 & x20 & x22 \\ x32 & x34 & x36 & x38 \\ x48 & x50 & x52 & x54 \end{bmatrix}$, $C'_{8x4} = C_{8x8}B_{8x4}$, and the subscriptions indicate the sizes of the matrixes.

So the matrix sizes during computation can be reduced and the times of multiplication and addition are also reduced. In hardware, it is realized by skipping part of the calculations for background type blocks. Similarly, during column DCT, part of the calculation can also be skipped since half of the inputs in each column are the same.

The analog data are read out from imager row by row. After ADC, the data needs to be stored in a memory temporarily for reordering before performing DCT. If the block type prediction is done during image processing, since the digital data is read out from the memory one by one, at least 64 additional clocks are needed in order to make a prediction for an 8×8 block. Also, additional memory is required to keep the data during prediction before DCT is performed. However, in our case, since the block type is decided in the imager during the image capture stage rather than during the image processing stage, it does not need additional processing time. Also the idea of adaptive complexity 2D-DCT can be combined with the dual analog Vdd algorithm proposed in [8], which makes the best of the block type decision circuitry.

3. Implementation of the Low Power Imaging System and Results

3.1. Image Sensor for Smart Image Capture with Block Type Decision Block

The proposed imager for smart image capture is mainly composed of a pixel array, row and column decoders, block decision block, and readout circuits, as shown in Figure 4. It is similar to the imager proposed in [8] but not exactly the same.

It works in a rolling shutter mode, the signals do not need to be sampled to in-pixel capacitors as required in [8] but to column sample and hold circuit. The readout and the decision operations share the same row select signals and row select transistors. However in [8], the imager works in a global shutter way and the decisions are made at the middle of the integration time, therefore the read out and the decision have separate row select signals and row select transistors. The pixel circuitry here is much simpler than that in [8].

The conventional 3T CMOS APS pixel is used in the pixel array [1]. A p-channel source follower is used to compensate for threshold voltage level-shifting from the n-channel, pixel-level source follower.

Figure 4. Block diagram of the proposed imager for smart image capture.

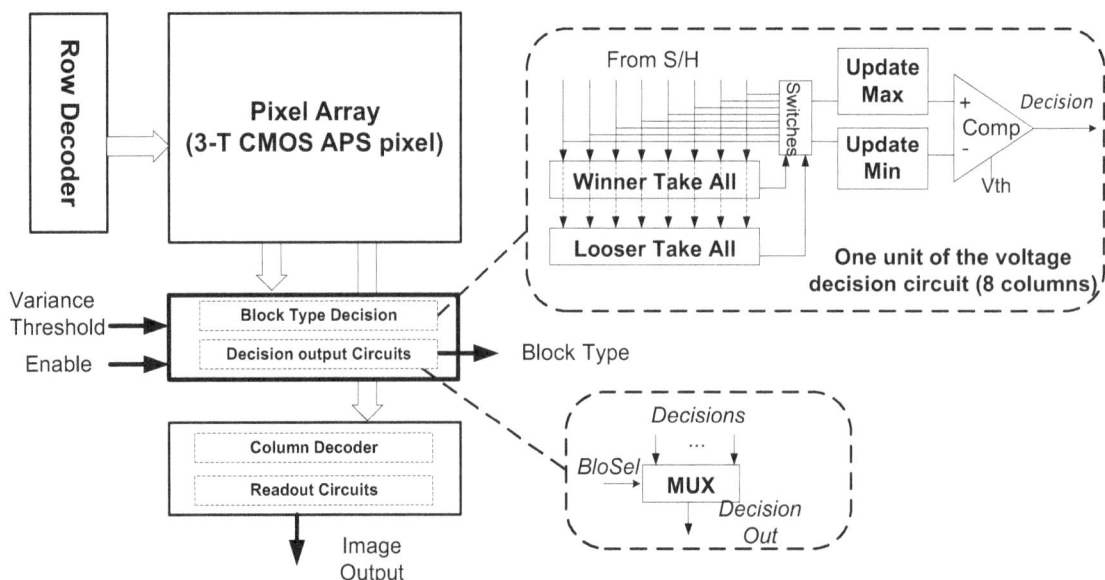

The block type decision unit is shared by each 8 columns and computes the type for each block according to the estimated voltage variance values $V_{Max} - V_{Min}$, similar to what was done in [8]. The enable signal activates the computations only when needed to save power. The block type decision unit is shown in Figure 4. Detailed description about Winner Take All (WTA), Loser Take All (LTA), Update Max/Min circuitries can be found in [8]. During the readout, the decision signal for each 8×8 block is output through a multiplexer one by one to the compression module for adaptive complexity controlling.

A chip for image capture and block type prediction is implemented based on TSMC 0.18 μm process, as shown in Figure 5. Its attributes are given in Table 1. The simulations are done by Cadence Spectre.

Figure 5. Layout of the image sensor with block type decision circuitry.

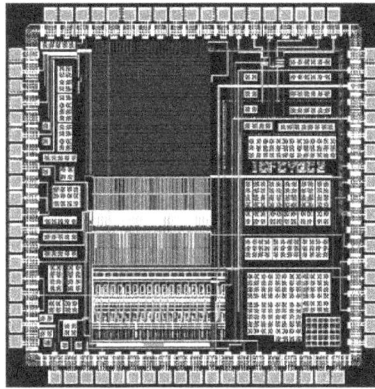

Table 1. Chip Attributes.

Technology	TSMC 0.18 μm
Voltage supply	1.8 V
Pixel array size	128 × 128
Pitch width	5 μm
Chip size	2 mm × 2 mm
Fill factor	26%
Estimated power (whole chip)	0.5 mW @ 30 FPS
Estimated power (decision logic)	7 μW @ 30 FPS

3.2. Image Sensor for Smart Image Capture with Block Type Decision Block Adaptive Complexity 2D-DCT Calculation

2D DCT can be done by running a 1D DCT over every row and then every column. Vector processing using parallel multipliers is a method used for implementation of DCT. The advantages of vector processing method are regular structure, simple control and interconnect and good balance between performance and complexity of implementation. The complexity of the 2D-DCT depends on the block type decided by the image sensor during image capture. Two optimizations are performed for the small variance blocks to save power.

3.2.1. Adaptive Data Format

For small variance blocks, only the differential part of the pixel values $V_{pixel} - V_{DC}$ are used for AC coefficients computing. Here, the VDC is the minimum value in each corresponding 8×8 block. In order to simplify the implementation and reduce the hardware requirement, the first pixel is used as the DC part instead of the minimum pixel in the block.

Figure 6 shows the circuit for implementing the adaptive input format according to block types. For large variance blocks, the pixel values are put for DCT calculation directly. For small variance blocks, the inputs are calculated by subtracting the first pixel value of a block V_{first_b} from the pixel values, and then input to the DCT block. The DC values should be compensated to the DC coefficients to generate the final DC coefficients. Another benefit brought by this optimization is a small increment in image quality of the reconstructed picture. The reason is that because fewer bits are performed for DCT coefficients calculation, less information is lost during the truncation stage. There are less values toggling during DCT coefficient calculation and therefore less power is consumed.

Figure 6. Schematic of adaptive input format according to block types.

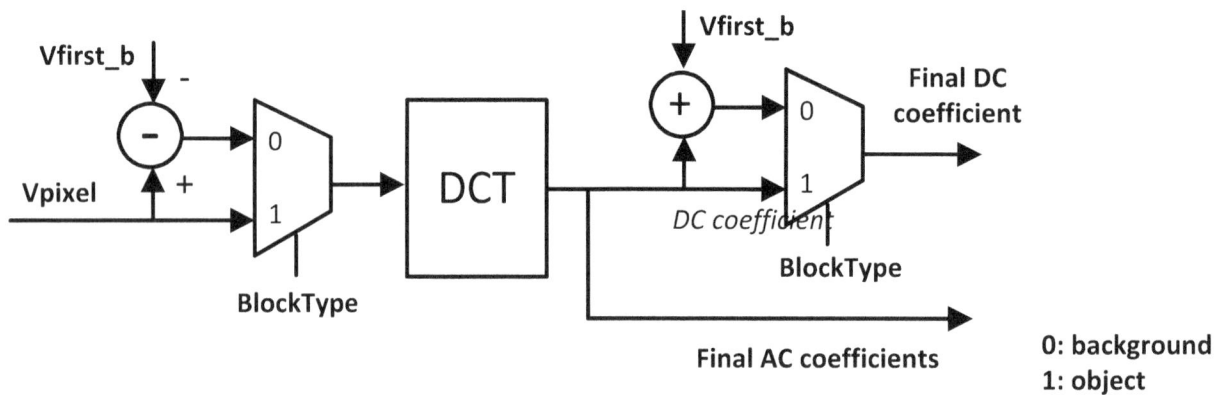

3.2.2. Adaptive Spatial Resolution

For small variance blocks, the spatial resolution for these blocks can be reduced while not affecting the image quality much. Consequently part of the calculations can be skipped to save power consumption during DCT. For small variance blocks, the calculation for the second row is just the same as the first row, therefore we can skip the row DCT alternatively. In addition, since half of the inputs of the non-skipped row DCT and column DCT are the same, half of the calculation groups can be skipped during calculation for small variance blocks, as shown in Figure 7.

For now, the adaptive complexity 2D-DCT is implemented based on FPGA (Cyclone EP1C20F400C8) first. Later we are planning to integrate the imager for capture and the compression on the same chip. According to the synthesis report given by Quartus II, there is about 10% hardware increase than a conventional 2D-DCT with unique complexity. The maximum frequency is 100 MHz.

Figure 7. Schematic of adaptive spatial resolution according to block types. (a) Normal case—large variance blocks; (b) Low-complexity case (dashed paths are disabled)—Small variance blocks.

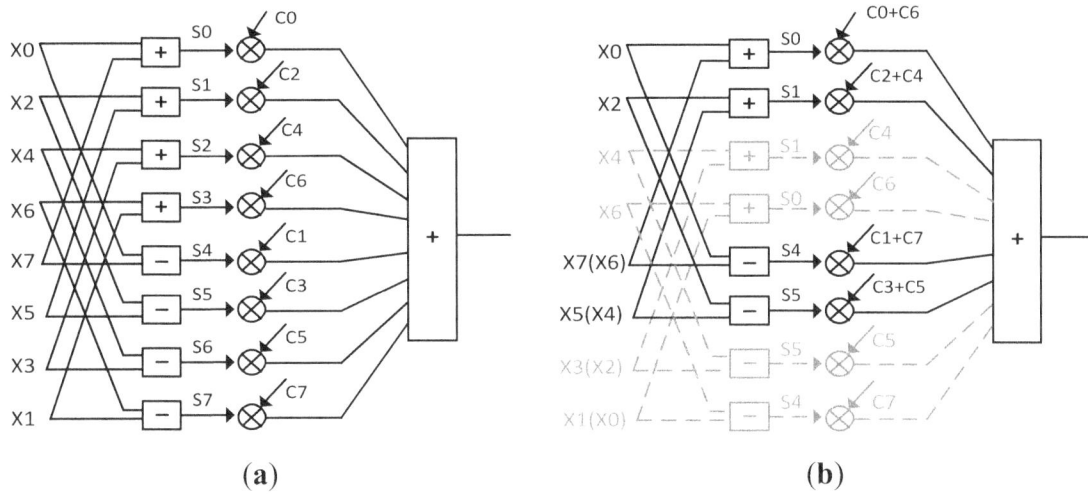

(a)   (b)

3.3. Performance

In order to show the benefit of the proposed low power algorithm, the proposed 2D-DCT based computation and a reference 2D-DCT core released by Xilinx [16] are implemented and compared.

As shown in Figure 8, three images of "Camera man", "Plane" and "Garden" which have background ratio of 50%, 90% and 0.8% small variance block ratios are used to represent three different types of images. The images have 8 bits resolution.

Figure 8. Test images (**a**) Camera man (background ratio is 50%); (**b**) Plane (background ratio is 90%); (**c**) Garden (background ratio is 0.8%).

(a)   (b)   (c)

Simulations about the PSNR *vs.* Variance threshold at different quantization levels are given in Figure 9a. At higher quantization level, the PSNR degradation is smaller. Therefore our low power algorithm is more efficient at higher quantization levels. The worst case happens when there is no quantization performed. The relationship between the Compression Ratio (CR) and variance threshold is given in Figure 9b. Since we have not applied entropy encoding yet, the Compression Ratio (CR) here is expressed by level of the quantization, that is, the percentage of the non-zero coefficients after quantization. The change of the compression ratio is not big since the compression is mainly done by DCT and quantization, the sub sampling by 2×2 on background blocks does not add much to the compression ratio. How PSNR changes with the variance threshold also depends on image types. Figure 10 shows the reconstructed image quality and power consumption for different types of images at different variance thresholds compared with those of a traditional compression in worst case (no quantization is performed).

Figure 9. (**a**) PSNR of reconstructed images *vs*. Variance threshold at different quantization levels; (**b**) Compression ratio and PSNR *vs*. Variance threshold.

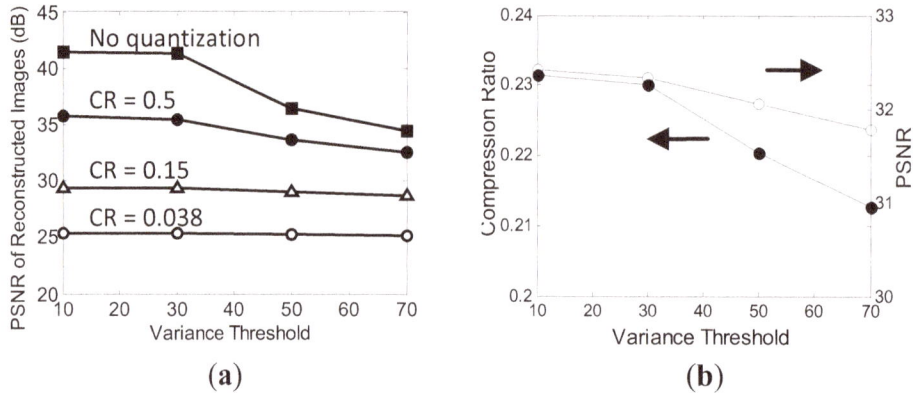

(a)   (b)

Figure 10. (**a**) PSNR and Power *vs*. Variance threshold for normal background images—Cameraman; (**b**) PSNR and Power *vs*. Variance threshold for flat background images—Plane; (**c**) PSNR and Power *vs*. Variance threshold for busy background images—Garden.

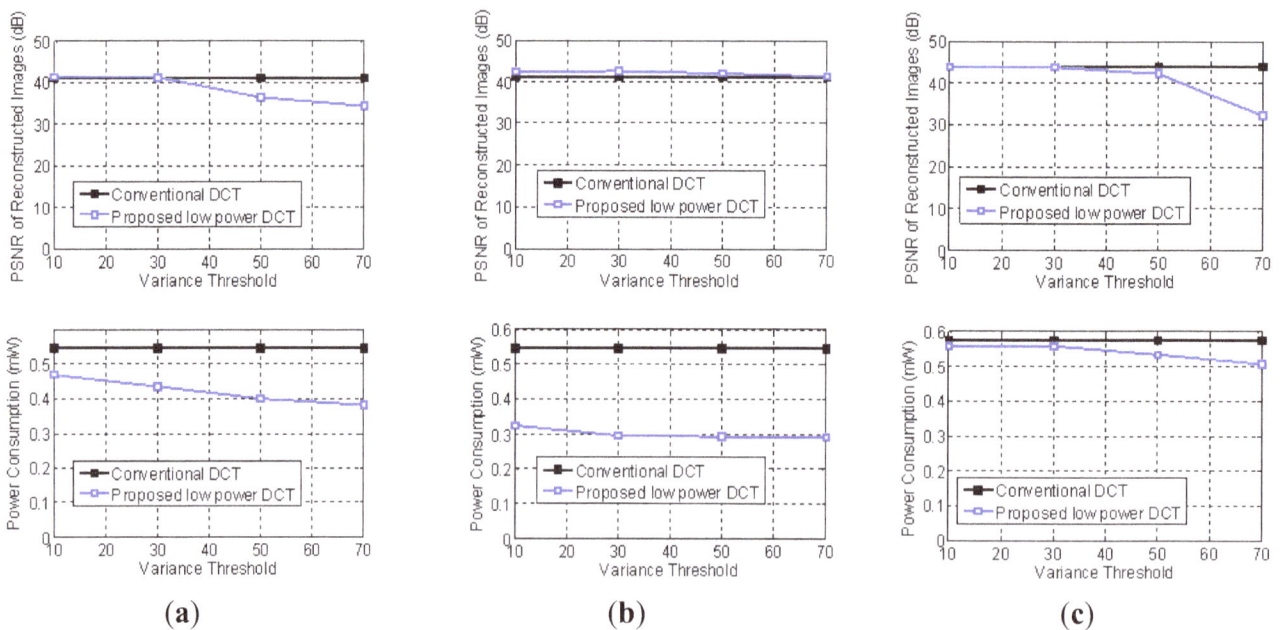

(a)   (b)   (c)

The power is estimated by Quartus II based on the Voltage Change Dump (VCD) files from post layout netlist simulation and the PSNR analysis is done by Matlab. The clock frequency used here is 0.5 MHz corresponding to a 128×128 array working at 30 FPS. It can be increased up to 100 MHz for imagers with larger array size and higher frame rate. The power savings varies from image to image. It is more efficient for predominant background images with up to about 46% power saving while no extra image quality degradation is observed compared with traditional compression. Extra image quality degradation is small because optimizations are performed only for background blocks. Power saving and the image quality of the reconstructed picture depend on the variance threshold. It is a tradeoff between image quality and power, and can be easily controlled by the threshold according to different applications. As shown in Table 2, by choosing the appropriate variance threshold, *i.e.*, 30 out of 256 for 8-bit resolution images, significant amounts of power can be saved with no extra image quality degradation.

Table 2. Power saved During 2D-DCT Calculation.

Images	Background ratio (th = 30)	Percentage of power saving	Extra image quality degradation
Garden	0.8%	0.5%	None
Cameraman	50%	24%	None
Plane	90%	46%	None

The power of the whole imaging system is the sum of the image sensor, ADC and the compression. For the proposed image sensor, the power estimation of the block type prediction is 7.0 μW. This adds only about 1.4% to the total imager power, and 0.7% to the system. Therefore, we conclude that the expected power savings outperform the extra power caused by the block type decision circuitry and result in significant power savings for the system.

4. Conclusions

A novel low power CMOS imaging system with smart image capture and adaptive complexity 2D-DCT calculation is proposed, simulated and implemented. The complexity of the 2D-DCT calculation is controlled by the block types which are estimated during image capture stage. It does not add additional processing time or memory space for block type prediction.

The imager is more efficient when the picture has predominant background. By choosing appropriate threshold, up to 46% of the power consumption can be saved during 2D-DCT calculation for images having predominant background, while no extra image quality degradation occurs for the reconstructed pictures compared with traditional compressions. For typical scenarios, up to about 23% of power can be saved for the whole imaging system. The idea of smart image capture and adaptive complexity according to block types can be extended to other 2D DCT architectures.

Acknowledgments

The authors would like to thank CMC for the support with Cadence tools and chip fabrication.

Conflict of Interest

The authors declare no conflict of interest.

References

1. Yadid-Pecht, O.; Etienne-Cummings, R. *CMOS Imagers: From Phototransduction to Image Processing*; Kluwer: Norwell, MA, USA, 2004.

2. Fish, A.; Yadid-Pecht, O. Low-Power "Smart" CMOS Image Sensors. In Proceedings of the IEEE International Symposium on Circuits and Systems, Washington, DC, USA, 18–21 May 2008; pp. 1408–1411.

3. Xu, C.; Zhang, W.; Ki, W. A 1.0 V VDD CMOS active-pixel sensor with complementary pixel architecture and pulsewidth modulation fabricated with a 0.25 μm CMOS process. *IEEE J. Solid-State Circuits* **2002**, *37*, 1853–1859.

4. Shen, C.; Xu, C.; Huang, R.; Zhang, W.; Ko, P.K.; Chan, M. A New APS Architecture on SOI Substrate for Low Voltage Operation. In Proceedings of the 9th International Symposium on IC Technology, Systems & Applications, Singapore, 3–5 September 2001; pp. 275–278.

5. Fish, A.; Hamami, S.; Yadid-Pecht, O. CMOS image sensors with self-powered generation capability. *IEEE Trans. Circuits Syst. II* **2006**, *53*, 131–135.

6. Shi, C.; Law, M.K.; Bermak, A. A novel asynchronous pixel for an energy harvesting CMOS image sensor. *IEEE Trans. VLSI Syst.* **2011**, *19*, 118–129.

7. Gao, Q.; Yadid-Pecht, O. Dual VDD Block Based CMOS Image Sensor–Preliminary Evaluation. In Proceedings of the IEEE International Symposium on Circuits and Systems (ISCAS), Rio de Janeiro, Brazil, 15–19 May 2011; pp. 1820–1823.

8. Gao, Q.; Yadid-Pecht, O. A low power block based CMOS image sensor with dual VDD. *IEEE Sens. J.* **2012**, *12*, 747–755.

9. Loeffler, C.; Lightenberg, A.; Moschytz, G.S. Practical Fast 1-D DCT Algorithms with 11-Multiplications. In Proceedings of the International Conference on Acoustics, Speech, and Signal Processing-89, Glasgow, UK, 23–26 May 1989; Volume 2, pp. 988–991.

10. Thoudam, V.P.S.; Bhaumik, B.; Chatterjee, S. Ultra Low Power Implementation of 2-D DCT for Image/Video Compression. In Proceedings of International Conference on Computer Applications & Industrial Electronics (ICCAIE 2010), Kuala Lumpur, Malaysia, 5–7 December 2010; pp. 532–536.

11. Sun, C.-C.; Ruan, S.-J.; Heyne, B.; Goetze, J. Low-power and high-quality Cordic-based Loeffler DCT for signal processing. *Circuits Devices Syst. IET* **2007**, *1*, 453–461.

12. Tran, T.D. The binDCT: Fast multiplierless approximation of the DCT. *IEEE Signal Process. Lett.* **2000**, *7*, 141–144.

13. Sung, T.-Y.; Shieh, Y.-S.; Yu, C.-W.; Hsin, H.-C. High-Efficiency and Low-Power Architectures for 2-D DCT and IDCT Based on CORDIC Rotation. In Proceedings of the 7th International Conference on Parallel and Distributed Computing, Applications and Technologies, Taipei, Taiwan, 4–7 December 2006; pp. 191–196.

14. Jeong, H.; Kim, J.; Cho, W.-K. Low-power multiplierless DCT architecture using image data correlation. *IEEE Trans. Consum. Electron.* **2004**, *50*, 262–267.

15. Xanthopoulos, T.; Chandrakasan, A.P. A low-power DCT core using adaptive bitwidth and arithmetic activity exploiting signal correlations and quantization. *IEEE J. Solid-State Circuits* **2000**, *35*, 740–750.

16. Pillai, L. *Video Compression Using DCT*; Xilinx: San Jose, CA, USA, 2002.

17. Loeve, M. *Probability Theory*, *Graduate Texts in Mathematics*, 4th ed.; Springer-Verlag: Berlin, Germany, 1977.

6

Hardware Implementation of an Automatic Rendering Tone Mapping Algorithm for a Wide Dynamic Range Display

Chika Ofili [1], **Stanislav Glozman** [2] **and Orly Yadid-Pecht** [1,*]

[1] Integrated Sensors, Intelligent Systems (ISIS) Laboratory, Electrical and Computer Engineering Department, University of Calgary, Calgary, AB T2N 1N4, Canada; E-Mail: chikaofili@yahoo.com
[2] VLSI Systems Center, Ben-Gurion University of the Negev, POB 653, Beer-Sheva 84105, Israel; E-Mail: stanislav.glozman@gmail.com

* Author to whom correspondence should be addressed; E-Mail: orly.yadid.pecht@ucalgary.ca

Abstract: Tone mapping algorithms are used to adapt captured wide dynamic range (WDR) scenes to the limited dynamic range of available display devices. Although there are several tone mapping algorithms available, most of them require manual tuning of their rendering parameters. In addition, the high complexities of some of these algorithms make it difficult to implement efficient real-time hardware systems. In this work, a real-time hardware implementation of an exponent-based tone mapping algorithm is presented. The algorithm performs a mixture of both global and local compression on colored WDR images. An automatic parameter selector has been proposed for the tone mapping algorithm in order to achieve good tone-mapped images without manual reconfiguration of the algorithm for each WDR image. Both algorithms are described in Verilog and synthesized for a field programmable gate array (FPGA). The hardware architecture employs a combination of parallelism and system pipelining, so as to achieve a high performance in power consumption, hardware resources usage and processing speed. Results show that the hardware architecture produces images of good visual quality that can be compared to software-based tone mapping algorithms. High peak signal-to-noise ratio (PSNR) and structural similarity (SSIM) scores were obtained when the results were compared with output images obtained from software simulations using MATLAB.

Keywords: tone mapping; wide dynamic range (WDR); high dynamic range (HDR); CMOS image sensors; hardware implementation; field programmable gate array (FPGA)

1. Introduction

Dynamic range can be described as the luminance ratio between the brightest and darkest part of a scene [1,2]. Natural scenes can have a wide dynamic range (WDR) of five (or more) orders of magnitude, while commonly used conventional display devices, such as a standard LCD display have a very limited dynamic range of two orders of magnitude (eight-bit, which can represent 256 levels of radiance).

Wide dynamic range (WDR) images, which are also called high dynamic range images (HDR) in the literature, can be obtained by using a software program to combine multi-exposure images [3]. Due to recent technological improvements, modern image sensors can also capture WDR images that accurately describe real world scenery [4–8]. However, image details can be lost when these captured WDR images are reproduced by standard display devices that have a low dynamic range, such as a simple monitor. Consequently, the captured scene images on such a display will either appear over-exposed in the bright areas or under-exposed in the dark areas. In order to prevent the loss of image details due to differences in the dynamic range between the input WDR image and the output display device, a tone mapping algorithm is required. A tone mapping algorithm is used to compress the captured wide dynamic range scenes to the low dynamic range devices with a minimal loss in image quality. This is particularly important in applications, such as video surveillance systems, consumer imaging electronics (TVs, PCs, mobile phones and digital cameras), medical imaging and other areas where WDR images are used in obtaining more detailed information about a scene. The proper use of a tone mapping algorithm ensures that image details are preserved in both extreme illumination levels.

Several advancements have occurred in the development of tone mapping algorithms [9]. Two main categories of tone mapping algorithms exist: tone reproduction curves (TRC) [10–14] and tone reproduction operators (TRO) [15–21].

The tone reproduction curve, which is also known as the global tone mapping operator, maps all the image pixel values to a display value without taking into consideration the spatial location of the pixel in question [21]. As a result, one input pixel value results in only one output pixel value.

On the other hand, the tone reproduction operator, also called the local mapping operator, depends on the spatial location of the pixel, and varying transformations are applied to each pixel depending on its surroundings [20,22]. For this reason, one input WDR pixel value may result in different compressed output values. Most tone mapping operators perform only one type of tone mapping operation (global or local); algorithms that use both global and local operators have been introduced by Meylan *et al.* [21,23], Glozman *et al.* [22,24] and Ureña *et al.* [25].

There are trade-offs that occur depending on which method of tone mapping is utilized. Global tone mapping algorithms generally have lower time consumption and computational effort, in comparison to local tone mapping operators [22]. However, they can result in the loss of local contrast, due to the global compression of the dynamic range. Local tone mapping operators produce higher quality images, because they preserve the image details and local contrast. Although local tone mapping algorithms do not result in a loss of local contrast, they may introduce artifacts, such a halos, to the resulting compressed image [22,24]. In addition, most tone mapping algorithms require manual tuning of their parameters in order to produce a good quality tone-mapped image. This is because the rendering parameters greatly affect the results of the tone-mapped image. To avoid adjusting the system's configuration for each WDR image, some developed tone mapping algorithms select a constant value for their parameters [21,26]. This makes the tone mapping operator less suitable for hardware video applications, where a variety of WDR images with different image statistics will be compressed. An ideal tone mapping system should be able to compress an assortment of WDR images while retaining local details of the image and without re-calibration of the system for each WDR image frame.

The tone mapping algorithm developed by Glozman et al. [22,24] takes advantage of the strengths of both global and local tone mapping methods, so as to achieve the goal of compressing a wide range of pixel values into a smaller range that is suitable for display devices with less heavy computational effort. The simplicity of this tone mapping operator makes it a suitable algorithm that can be used as part of a system-on-a-chip. Although it is an effective tone mapping operator, it lacks a parameter decision block, thereby requiring manual tuning of its rendering parameters for each WDR image that needs to be displayed on a low dynamic range device.

In this paper, a new parameter setting method for Glozman et al.'s operator and a hardware implementation of both the estimation block and the tone mapping operator is proposed [22,24]. The novel automatic parameter decision algorithm avoids the problem of manually setting the configuration of the tone mapping system in order to produce good quality tone-mapped images. The optimized tone mapping architecture was implemented using Verilog Hardware Description Language (HDL) and synthesized into a field programmable gate array (FPGA). The hardware architecture combines pipelining and parallel processing to achieve significant efficiency in processing time, hardware resources and power consumption.

The hardware design provides a solution for implementing the tone mapping algorithm as part of a system-on-a-chip (SOC) with WDR imaging capabilities for biomedical, mobile, automotive and security surveillance applications [1,27]. The remainder of this paper is structured as follows: Section 2 provides a review of existing tone mapping hardware architectures. Section 3 describes the tone mapping algorithm of Glozman et al. [22,24]. The details of the proposed extension of the tone mapping operator are described in Section 4. The results of the proposed optimization of the tone mapping operator are also displayed in Section 4. Section 5 explains the proposed hardware architecture in detail. In Section 6, the hardware cost and synthesis details of the full tone mapping system are given. Comparisons with other published hardware designs in terms of image quality, processing speed and hardware cost are also presented in Section 6. In addition, simulation results of the automated hardware implementation using various WDR images obtained from the Debevec library and other sources are shown in Section 6. Finally, conclusions are presented in Section 7.

2. Related Works on Tone Mapping Systems

2.1. Tone Mapping Implementations: Hardware vs. Software

Tone mapping algorithms can be executed on either hardware or software systems or on an incorporation of both. A high number of successful tone mapping algorithms have been implemented on software platforms [12,17–21,26,28–32]. For software implementations, the tone mapping operator is applied on the stored image using a programmed application on a PC workstation. Different tone mapping algorithms with varying complexities can be easily developed in software. Although this is advantageous, because modifications to the software code can be made quickly, complications may arise when trying to implement a tone mapping system with high processing speed for WDR video technology applications. Software systems can achieve real-time implementation for global tone mapping systems. Although global tone mapping algorithms have a shorter computational time, they may result in a loss of local contrast in WDR images. In contrast, complications may arise when implementing local tone mapping algorithms, due to computational complexity and the time required for these algorithms to process one image frame [29]. In addition, software implementation of a tone mapping system will not satisfy our goal of having the architecture embedded as part of a system on chip.

There are a lot of hardware-based platforms, such as graphic processing units (GPUs), field programmable gate arrays (FPGAs), application-specific integrated circuits (ASICs) etc., that can be used to achieve real-time operation for both global and local tone mapping systems. Hardware-based implementations have an added advantage over software developed systems in terms of the ability of embedding the tone mapping system as part of a portable camera device for applications, such as consumer imaging electronics (TVs, PCs, mobile phones and digital cameras), medical imaging, video surveillance systems etc. Captured images can be processed pixel-by-pixel within the WDR image sensor silicon without the need of external memory buffers, thereby reducing the amount of power consumed and the area occupied by the tone mapping operators.

The tone mapping hardware system discussed in this paper was developed for a low cost FPGA device as a proof of concept. It can be further implemented in ASIC as part of a system-on-a-chip with a WDR image sensor for high volume production. In addition, a comparison with other FPGA-based tone mapping systems available in terms of hardware resources, processing speed and power consumption is presented. It should be noted that no hardware-based design of an exponent-based tone mapping operator [22,24] has been implemented and no automatic parameter decision algorithm has been proposed for it, either. This is the main thrust of this work.

2.2. Related Research on Hardware-Based Tone Mapping

There have been a few hardware-based implementations of tone mapping operators. An FPGA-based architecture by Hassan et al. [33] implements a modification of the local tone mapping operator by Reinhard et al. [18]. Modifications were made to the original local tone mapping operator in order to reduce the hardware complexity of the algorithm. It was implemented on an Altera® Stratix II FPGA operating at 60 frames per second for a 1024 × 768 pixel image. A patent was filed for the above hardware implementation with an extension of the algorithm for colored WDR images [34].

Chiu et al. [35] presented an application-specific integrated circuit (ASIC) implementation of both the global tone mapping operator by Reinhard et al. [18] and gradient compression by Fattal et al. [26] using an ARM-SOC platform. The tone mapping processor can compress a 1024 × 768 image at 60 frames per second (fps) with an operating frequency of 100 MHz. It also occupies a core area of 8.1 mm² under TSMC 0.13 μm technology.

Vytla *et al.* [36] implemented another real-time hardware tone mapping algorithm based on the gradient domain dynamic range compression by Fattal *et al.* [26]. A modified Poisson solver is utilized in order to reduce the hardware complexity of the original Poisson equation. A processing speed of 100 fps at a maximum operating frequency of 114.18 MHz was achieved for a one megapixel frame using the Stratix II FPGA.

A new global and local tone mapping algorithm by Ureña *et al.* [25] was designed and implemented on a GPU NVIDIA ION2 and Xilinx® Spartan III FPGA for VGA resolution WDR images (640 × 480). The tone mapping operator can adapt VGA images at 30 fps and 60 fps for the GPU and FPGA device, respectively. In addition, a power consumption of 900 mV was achieved for an FPGA-based system with the same image resolution [37].

Kiser *et al.* [38] recently implemented on a Xilinx® Spartan VI FPGA the global tone mapping algorithm by Reinhard *et al.* [18], as well as its automatic parameter estimation method for tuning one of the rendering parameters [39].

In general, the hardware-based designs presented so far can compress WDR images in real-time using the different tone mapping algorithms implemented. In this paper, however, we focused on designing an efficient hardware tone mapping system by optimizing the tone mapping algorithm of Glozman *et al.* [22,24] and implementing the improved tone mapping design in an FPGA device that will be used along with a WDR CMOS imager for real-time image compression. The motivation here is to later integrate the tone mapping system within a single silicon as part of a system-on-a-chip with WDR image capture capabilities. Our tone mapping system requires fewer hardware resources, lower power consumption and has higher processing speed when compared to other's work in the past. This is achieved by using a combination of parallel processing along with pipelining to achieve significant efficiency in processing time, hardware resources and power consumption.

3. Tone Mapping Algorithm for Current Work

The tone mapping algorithm used in this work was originally designed by Glozman *et al.* [22,24]. It performs a mixture of both global and local compression on colored WDR images. The aim of Glozman *et al.* [22,24] was to produce a tone mapping algorithm that does not require heavy computational effort and can be implemented as part of a system on a chip with a WDR image sensor. In order to achieve this, they obtained inspiration from the approach proposed by Meylan *et al.* [21] that involves directly implementing the tone mapping operator on the Bayer color filter array (CFA). This is different from other traditional tone mapping systems that apply the tone mapping algorithm after demosaicing the Bayer CFA image. A visual comparison of the two different color image processing workflow is shown in Figure 1. In traditional color image processing workflow, the image is first converted to YUV or HSV color space (luminance and chrominance), and the tone mapping algorithm is then applied on the luminance band of the WDR image [18,25,26]. The chrominance layer or the three image color components (R, G, B) are stored for later restoration of the color image [26]. This is not ideal for a tone mapping operator that will be implemented solely on a system-on-a-chip because additional memory is needed to store the full chrominance layer or the RGB format of the image. In contrast, no extra memory is required to store the chrominance layer of the colored image being processed in

Glozman *et al.*'s tone mapping approach, since it uses the monochrome Bayer image formed at the image sensor [22,24]. Furthermore, no additional computation is required for calculating the luminance layer of the colored image, thereby reducing the overall computational complexity of our algorithm.

Figure 1. (**a**) Workflow approach proposed by Meylan *et al.* [21] and implemented in Glozman *et al.*'s tone mapping algorithm [22,24]; (**b**) Detailed block description of a traditional color image processing workflow for tone mapping wide dynamic range (WDR) images.

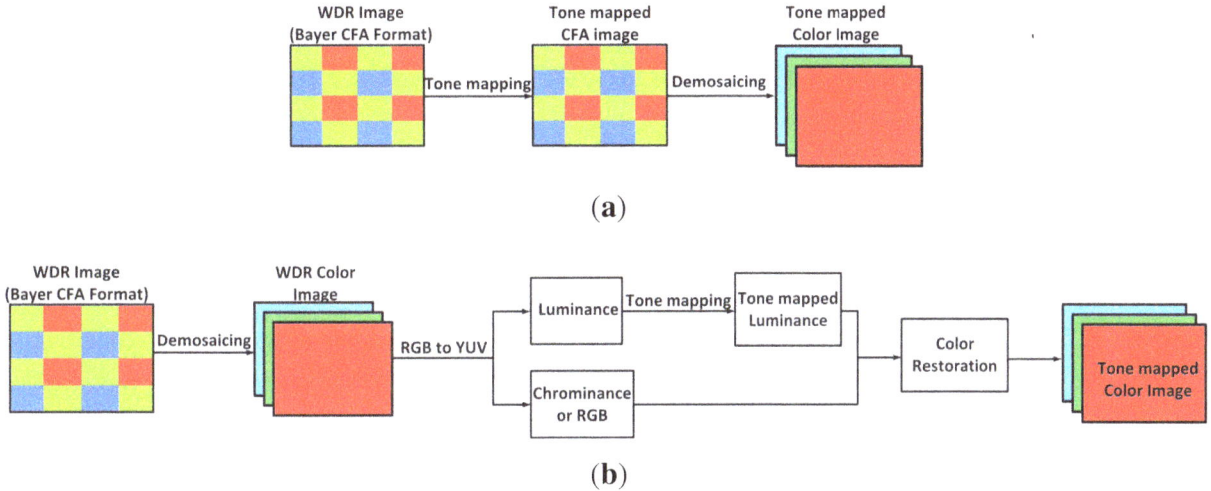

(**a**)

(**b**)

Glozman *et al.*'s tone mapping model takes inspiration from an inverse exponential function, which is one of the simplest nonlinear mappings used in adapting a WDR image to the limited dynamic range of display devices [22,24,29]. The inverse exponential function is defined as follows [29]:

$$Y(p) = 1 - \exp^{-\frac{X(p)}{X_o(p)}} \qquad (1)$$

where p is a pixel in the CFA image; $X(p)$ represents the input light intensity of pixel p; $Y(p)$ is the pixel's tone-mapped signal and $X_o(p)$ is the adaptation factor. $X_o(p)$ is a constant for every pixel, and it is given by the mean of the entire CFA image. In Glozman *et al.*'s tone mapping algorithm [22,24], the adaptation factor, $X_o(p)$, varies for each WDR pixel, and it is defined by:

$$X_o(p) = k \cdot X_{DC} + (X * F_w)(p) \qquad (2)$$

where p is a pixel in the image; $X_o(p)$ is the adaptation factor at pixel p; X_{DC} is the mean value of all the CFA image pixel intensities; * represents the convolution operation and F_w is a two-dimensional Gaussian filter. The output of the convolution is used in providing the local adaptation value for each CFA pixel, since that output depends on the neighborhood of the CFA pixel. The factor, k, is a coefficient, which is used in adjusting the amount of global component in the resulting tone-mapped image. Its value ranges from zero to one, depending on the overall tone of the image.

The new adaptation factor for the proposed tone mapping algorithm Equation (2) ensures that different tonal curves are applied to each pixel depending on the spatial location of the image pixel. This is different from the original inverse exponential tone mapping function that applies the same tonal curve to all the pixels in the WDR image.

A scaling coefficient is also added to Equation (1), so as to ensure that all the tone mapping curves will start at the origin and meet at the maximum luminance intensity of the display device. The full tone mapping function is defined as follows:

$$Y(p) = \frac{\alpha}{1 - \exp^{-\frac{X_{max}}{X_o(p)}}} \cdot \left(1 - exp^{-\frac{X(p)}{X_o(p)}} \right) \tag{3}$$

where X_{max} is the maximum input light intensity of the WDR image; X, and α is the maximum pixel intensity of the display device. For standard display devices, such as a computer monitor, $\alpha = 255$ (eight bits).

4. Proposed Automatic Parameter Estimation for the Exponent-Based Tone Mapping Operator

Glozman *et al.*'s operator is a tone mapping algorithm that compresses a WDR image based on its global and local characteristics [22,24]. It consists of global and local components that are used in the adaptation of a WDR image as given in Equations (2) and (3). The global component in adaptation factor $X_o(p)$ is comprised of factor k and the average of the image X_{DC}. The factor, k, is used in modulating the amount of global tonal correction X_{DC} that is done to the WDR image. It can be adjusted between zero and one, depending on the image key. The image key is a subjective quantification of the amount of light in a scene [18]. Low key images are images that have a mean intensity that is lower than average, while high key images have a mean intensity that is higher than average [40]. Analysis shows that low key WDR images require factor k values closer to zero, while factor k values closer to one are required for high key WDR images. This is because the lower the k value, the higher the overall image mean. On the other hand, the higher the k value, the greater the compression of higher pixel values, resulting in a less exposed tone-mapped image. To visually demonstrate the sole effect of the factor k on the tonal curve applied to the input WDR signal X, the local adaptation, $(X * F_w)(p)$, is kept constant, or it is assumed that there are no variations in the intensity of the surrounding pixels for each pixel in X, while the parameter k is used for performing tone mapping varies from 0.1 to one (Figure 2).

Figure 2. Inverse exponent curves for different k values [assuming $(X * F_w)(p)$ is constant for the given WDR input X].

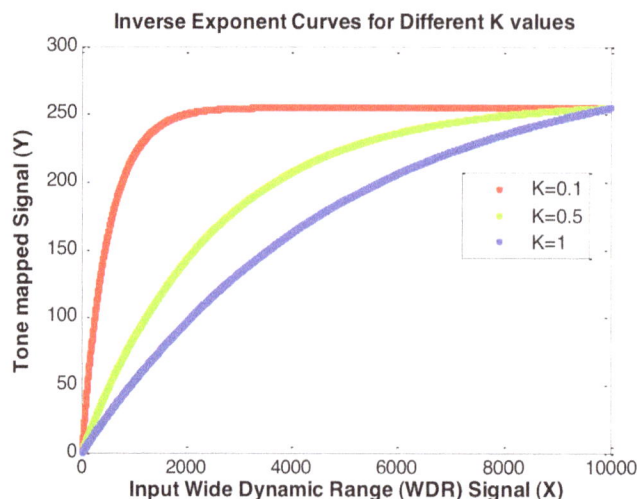

As shown in Figure 2, as the factor k decreases, more of the input WDR signal X is compressed to the higher ranges of the low dynamic range device (in this case, closer to 255). This results in an increase in the overall brightness of the image as factor k decreases. Thus, the rendering parameter k needs to be adjusted automatically in order to make the tone mapping system more compact for a system-on-a-chip application.

4.1. Developing the Automatic Parameter Estimation Algorithm

The aim was to design a simple hardware-friendly automatic parameter selector for estimating factor k, so that natural looking tone-mapped images can be produced using Glozman *et al.*'s tone mapping algorithm [22,24]. The proposed algorithm has to be hardware-friendly, so that it can be implemented as part of the tone mapping system without dramatically increasing the power consumption and hardware resources utilized. To derive this k-equation, a mathematical model based on the properties of the image needs to be obtained experimentally. The initial hypothesis is that there exists a one-dimensional interpolation function for computing the k value. Based on this hypothesis, the k-equation can be described as shown in Equation (4):

$$k = f(x) \tag{4}$$

where k is the dependent variable that is applied with the adaptation equation [Equation (2)], which varies from zero to one, and x is an independent property of the WDR image. Since the function needs to be derived experimentally, the next step was to obtain the data samples used in modelling the algorithm. An outline of the steps taken in modeling the automatic estimation algorithm is shown in Figure 3. First, 200 WDR images were tone-mapped in order to obtain good quality display outputs, and their respective k values were stored. Then, the same WDR images were analyzed, and various independent variables were proposed and computed for each WDR image. Finally, the k values, as well as the different proposed independent variables were used in determining if a one-dimensional relationship exists between the selected k values and their corresponding independent variables. This is so that an equation can be derived for estimating the value k for producing good quality tone-mapped images.

Figure 3. Outline of the steps taken in deriving the mathematical model for estimating the k value needed for the tone mapping operation.

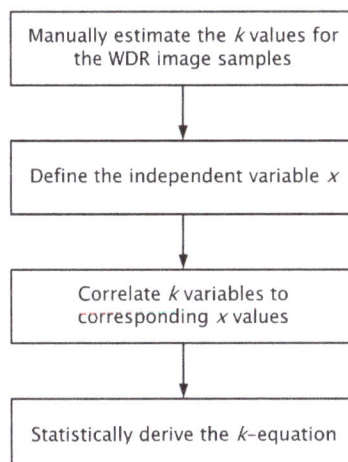

4.1.1. Manually Estimating the Values of k

In order to develop an automatic parameter estimation algorithm, a way of manually determining which values of k produced tone-mapped images that can be deemed natural looking had to be found. Previous studies on perceived naturalness and image quality have shown that the observed naturalness of an image correlates strongly to the characteristics of the image, primarily brightness, contrast and image details [41–43]. As previously highlighted, the factor k is a major parameter in the tone mapping algorithm, and its value may vary for different WDR images. Analysis performed showed that the value of k predominately affects the brightness level of the tone-mapped image, but also influences the contrast of the image, due to the increase or decrease of the image brightness, as shown in Figure 2. The amount of contrast and preserved image details is primarily determined by the local adaptation component of the tone mapping algorithm [22,24]. Since the goal here was to derive a simple one-dimensional interpolation function for computing factor k for the purpose of easy hardware implementation, our focus was mainly on selecting what level of brightness was good enough for a tone-mapped image. This resulted in the issue of how to quantitatively determine how bright a tone-mapped image should be in order for it to be categorized as being of good quality.

The statistical method by Jobson *et al.* [41] was used in estimating what the illumination range should be in order for an image to be perceived as being a visually pleasant image. In Jobson *et al.*'s work, they developed an approach of relating a visually pleasant image to a visually optimal region in terms of its image statistics, *i.e.*, image luminance and contrast. The overall image luminance is measured by calculating the mean of the image's luminance, μ, while the contrast σ is measured by calculating the global mean of the regional standard deviations. To obtain the regional standard deviations, non-overlapping area blocks of 50×50 pixel size were used to calculate the standard deviation of a pixel region in their experiment [41]. The size of the pixel regions for computing the standard deviation can be modified. Based on the results from their experiment, they concluded that a visually optimal image was found to have high image contrast, which lies within 40–80, while its image luminance is within 100–200 for an eight-bit image or, more specifically, around the midpoint of the images dynamic range [41]. This is to ensure that the image has a well-stretched histogram. Figure 4 shows the diagram of the visually optimal region for an image in terms of its image luminance and contrast [41].

Figure 4. The four image quality regions partitioned by image luminance and image contrast [41].

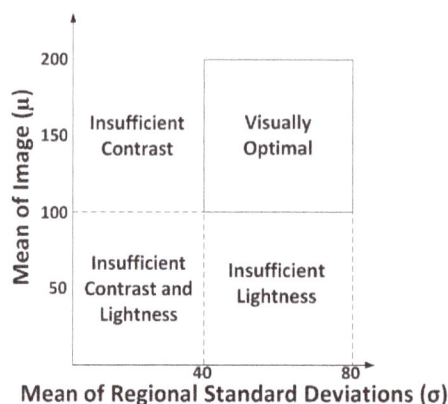

Using Jobson *et al.*'s theory, an experiment involving 200 WDR images was performed in order to obtain k values for the 200 tone-mapped images. In addition, the tone mapping image quality assessment tool by Yeganeh *et al.* [42] was used to ensure that the selected k values and the corresponding tone-mapped images were within the highest image quality scores attainable by the tone mapping algorithm. This objective assessment algorithm produces three image quality scores that are used in evaluating the image quality of a tone mapped image: the structural fidelity score (S), the image naturalness (N) score and the tone mapping quality index (TMQI). The structural fidelity score is based on the structural similarity (SSIM) index, and it calculates the structural similarity between the tone-mapped image and the original WDR image. The naturalness score is based on a statistical model that perceives that the naturalness of an image correlates with the image's brightness and contrast. The overall image quality measure, which the authors call the tone mapping quality index (TMQI), is computed as a non-linear combination of both the structural similarity score (S) and the naturalness score (N). The three scores generally range from zero to one, where one is the highest in terms of image quality. Figure 5 shows the histogram of the TMQI scores for the 200 WDR images. Since the tone mapping algorithm developed by Glozman *et al.* [22,24] is applied on a CFA image instead of the luminance layer, the calculation of the overall image mean and contrast was based on the mean of the CFA image, as well as the mean of the regional standard deviations, respectively.

Figure 5. Histogram of the tone mapping quality index (TMQI) scores of the 200 tone-mapped images.

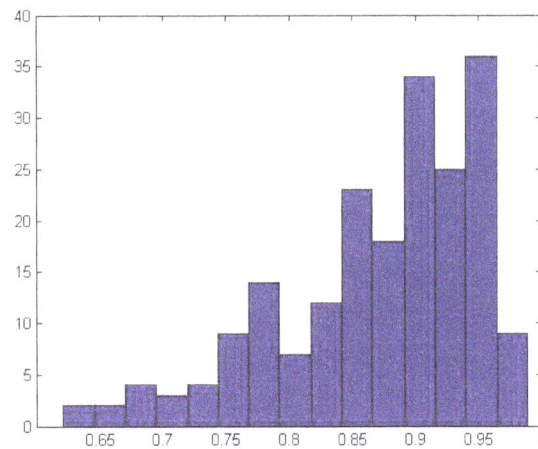

4.1.2. Determining the Independent Variable x

After performing the previous experiment where the "optimal" k values for 200 WDR images were obtained, the next step was to find a variable that correlates with these k values, so that the value of k can be predetermined before the actual tone mapping process.

An approach based on deriving an independent variable from an initially compressed WDR image was proposed. This is because WDR images can be seen as belonging to different classes, depending on the image's dynamic range, and this may affect the interpretation of the image's statistics. For example, a 10-bit WDR image (for which the intensity ranges from zero to $2^{10} - 1$), with a mean of 32, may be different from a 16-bit WDR image (zero to $2^{16} - 1$) with the same mean, due to the large difference in dynamic range. Since the goal here is to design a parameter estimation algorithm for Glozman *et al.*'s tone mapping operator that can be robust enough to estimate the k values for a large variety of WDR images (anything greater than eight bits), all the different WDR images need to be normalized to the same dynamic scale, so as to remove the effect of the different dynamic ranges.

The initial image compression stage is done using a simplified version of Glozman *et al.*'s tone mapping operation [Equation (3)]. Here, the scaling factor is removed, and the tone mapping algorithm is simplified to this:

$$Y(p) = \alpha \cdot \left(1 - \exp^{-\frac{X(p)}{X_o(p)}}\right) \tag{5}$$

$$X_o(p) = k \cdot X_{DC} + (X * F_w)(p) \tag{6}$$

where $Y(p)$ is the tone-mapped image used for estimation; α is the maximum pixel intensity of the display device and $X_o(p)$ is the adaptation factor, which is the same as that in Glozman *et al.*'s operator [Equation (2)]. For standard display devices, such as a computer monitor, $\alpha = 255$ (eight bits). For the adaptation function [Equation (6)], k was kept constant at 0.5, while the sigma and kernel size for the Gaussian filter was set to $\sigma = 3$ and 9×9, respectively.

Using the compressed image, four potential independent variables were investigated: minimum, maximum, mean and image contrast. These variables along with the WDR image's k values are compared in order to determine if there exists a correlation. The variables that have a one-dimensional relationship with the k value applied in the actual tone mapping operation are then used in statistically-deriving a k-equation. It should be noted that the same WDR images used in obtaining the 200 k values in the previous experiment were used for deriving the four potential variables.

4.1.3. Correlation between k and the Proposed Independent Variables

In order to derive an algorithm for estimating the values of k needed for tone mapping WDR images, the computed minimum, maximum, contrast and mean values of the 200 compressed images were analyzed. Results showed that there is a correlation between the mean (y_m) of the initial compressed image and the k value used in the actual tone mapping operation. As shown in Figure 6, the trend of the values of k selected for the varying images correlate with the description of the rendering parameter k. As expected, low k values are required for images with a lower-than-average mean intensity, while high k values are needed for WDR images with higher-than-average mean intensity. Thus, dark WDR images that appear as low key images require low k values for the actual tone mapping process, which, in fact, improves the brightness of the image, while the reverse occurs for extremely bright WDR images.

The results also show that there is no strong one-dimensional relationship between the tone-mapped image's contrast and the selected k value. Thus, in order to simplify the automatic parameter estimation model, the initial tone-mapped image contrast was not included in the derivation of the k-equation. Similar observations were found in the analysis of the minimum and maximum results. Hence, only the mean of the initially-compressed images was used in derivation of the estimation algorithm, and it will be used as the independent variable of the function [Equation (4)].

To obtain a relatively simple k-equation that can be implemented in hardware, the graph of k was divided into three regions (A, B, C). Region A represents extremely low independent variable y_m values; and the value of k in this region is kept constant [Equation (7)]. In Region C, the k value is also kept constant, and the region represents extremely high independent variable y_m values. Finally, Region B represents the region between the low and high range for the independent variable y_m. Using MATLAB®'s curve fitting tool, a k-equation (7) was obtained for Region B, which should be easy to implement in hardware.

$$k = \begin{cases} 0.035, & \text{if } y_m \leq 27.16 \\ 0.01439 \times 2^{0.04728 \cdot y_m}, & \text{if } 27.16 < y_m < 108.25 \\ 0.5, & \text{otherwise} \end{cases} \quad (7)$$

Figure 6. The graph of the mean of the initially-compressed image against the selected k for the 200 WDR images used in the experiment.

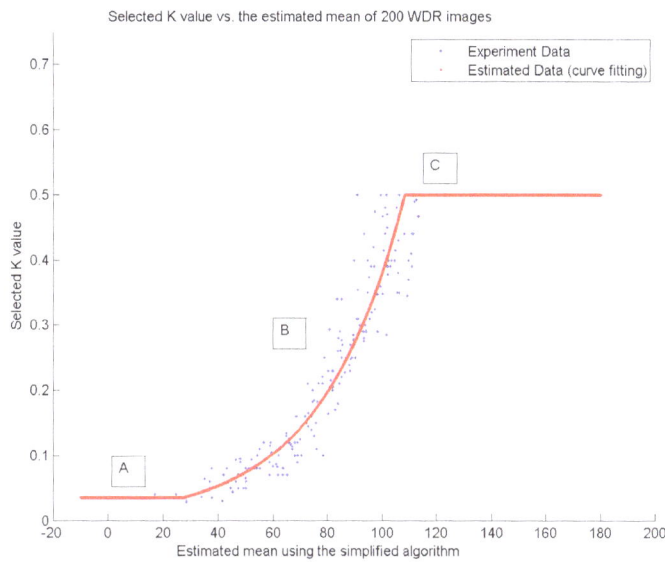

To evaluate the fit of the curve attained in Region B, the goodness of fit for the curve is computed. There are three criteria used in the analysis of these curves: the sum of squares due to error (SSE), R-squared and the root mean squared error (RMSE). The SSE is used to measure the total variation of the model's predicted values from the observed values [44]. A better fit is indicated by a score closer to zero. R-squared is used to measure how well the curve fit explains the variations in the experiment data [44,45]. It can range from zero to one, where a better fit is indicated by a score closer to one. RMSE is used to measure the difference between the model's predicted values and the experimentally-derived values [45]. A better fit is indicated by an RMSE value closer to zero. Results show that good SSE, RMSE and R-squared were attained for the fit. The SSE, RMSE and R-squared scores for the curve obtained in Region B were found to be 0.23300, 0.93160 and 0.03549, respectively.

Using the derived equation from the estimation model, a pathway for performing an automatic tone mapping on a WDR image was designed. This automatic pathway is shown in Figure 7. First, an estimation of the image statistics is performed by computing the mean of the initial compressed WDR image y_m using the simpler tone mapping algorithm mentioned above. Next, a k value is calculated using an estimation algorithm. This k value is then used for the actual tone mapping process using Glozman *et al.*'s operator [22,24].

Figure 7. Block diagram of the pathway used in performing automatic image compression of WDR images.

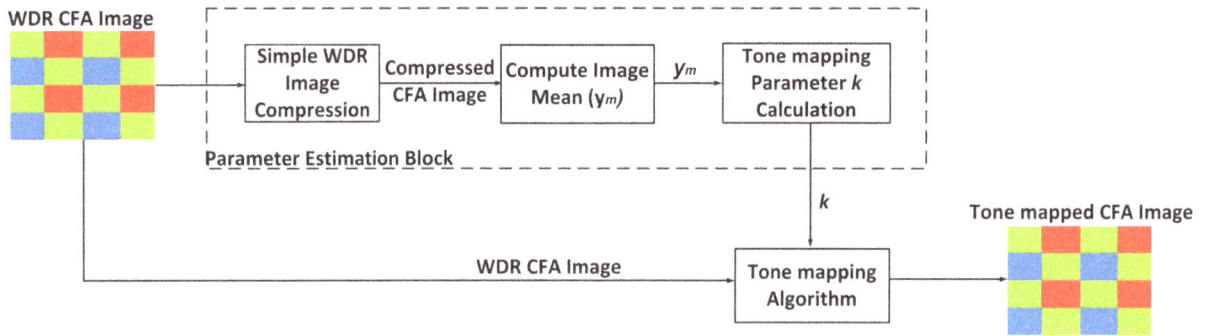

4.2. Assessment of the Proposed Extension of Glozman et al.'s Tone Mapping Operator

To demonstrate that the model [Equation (7)] produces satisfactory results for a variety of WDR images, the models were tested on images that were not used in deriving the equations. The objective assessment tool index by Yeganeh *et al.* [42] was used in selecting the k values for the manual implementation of Glozman et al.'s tone mapping operator. The k values that were used to produce images with the highest TMQI score attainable by the tone mapping algorithm were selected. Figures 8 and 9 show examples of the results from the automatic and manual-tuned tone mapping systems. Results show that the automatic rendering parameter model works well in estimating the parameter k. The image results produced using the automatic rendering parameter algorithm are similar to those obtained from manually tuning the tone mapping operator to conclude that the automatic rendering parameter model is suitable for compressing a variety of WDR images.

Figure 8. Tone mapped multi-fused image obtained using (**a**) Linear mapping; (**b**) Manually tuned Glozman *et al.*'s operator ($k = 0.270$) [22,24]; (**c**) The proposed automatic tone mapping algorithm ($k = 0.278$) [22,24]; (**d**) Meylan *et al.*'s operator [21]; (**e**) Reinhard *et al.*'s operator [18]; (**f**) The gradient-based tone mapping operator by Fattal *et al.* [26]; and (**g**) Drago *et al.*'s operator [12].

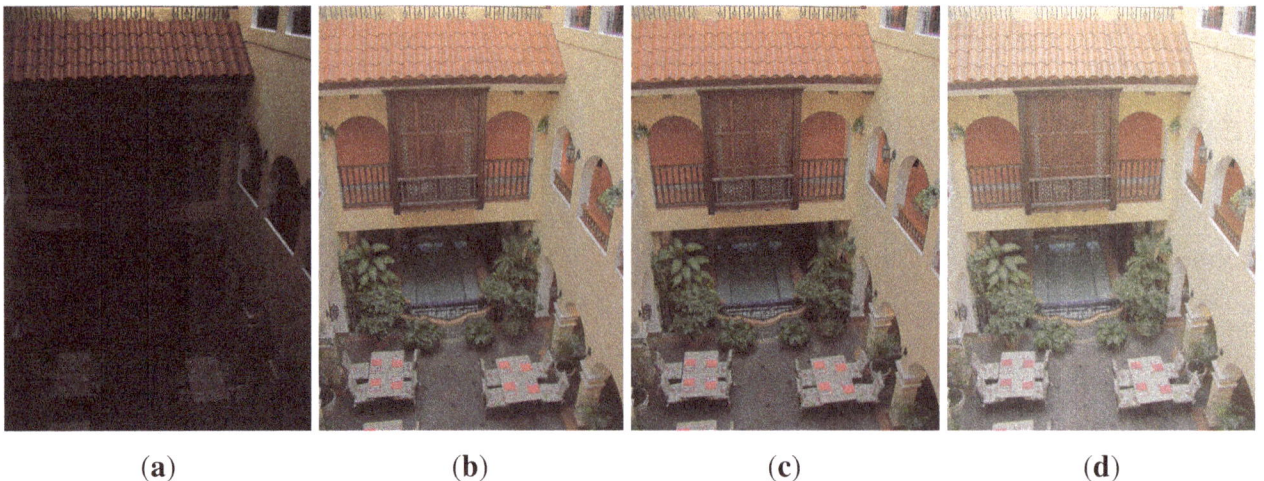

(**a**)   (**b**)   (**c**)   (**d**)

Figure 8. *Cont.*

(**e**)   (**f**)   (**g**)

Figure 9. Tone mapped multi-fused image obtained using (**a**) linear mapping; (**b**) manually tuned Glozman *et al.*'s operator ($k = 0.165$) [22,24]; (**c**) the proposed automatic tone mapping algorithm ($k = 0.177$) [22,24]; (**d**) Meylan *et al.*'s operator [21]; (**e**) Reinhard *et al.*'s operator [18]; (**f**) the gradient-based tone mapping operator by Fattal *et al.* [26]; and (**g**) Drago *et al.*'s operator [12].

(**a**)   (**b**)   (**c**)   (**d**)

(**e**)   (**f**)   (**g**)

Furthermore, a quantitative assessment was performed using the objective test developed by Yeganeh *et al.* [42], so as to compare the output images (manually tuned and automatically rendered) with other tone mapping operators. Images from tone mapping operators by Reinhard *et al.* [18], Fattal *et al.* [26] and Drago *et al.* [12] were obtained using the Windows application *Luminance HDR*, which is available at [46]. The MATLAB® implementation of Meylan *et al.*'s tone mapping algorithm [21] was also used in the comparison. The objective test score TMQI for the tone-mapped images shown in Figures 8 and 9 are displayed in Tables 1 and 2. Although Fattal *et al.*'s [26] gradient tone mapping operator had the highest structural similarity (S) scores, higher naturalness scores were attained by Glozman *et al.*'s operator (manual and automatic models) for the same test images. The results also show that image quality scores obtained for the automatic rendering model and manual calibration varied slightly. This indicates that the k-estimation method can produce "natural looking" tone-mapped images that are similar to those produced from tuning the rendering parameters manually.

Table 1. Quantitative measures for Figure 8.

Tone mapping operators	Structural fidelity	Naturalness	TMQI
Linear mapping	0.61635	0.00013	0.69174
Manual-tuned Glozman *et al.*'s operator [22,24]	0.85987	0.48439	0.88412
Proposed automatic tone mapping [22,24]	0.86126	0.48002	0.88373
Meylan *et al.* [21]	0.84081	0.15929	0.81405
Reinhard *et al.* [18]	0.89649	0.30586	0.86082
Fattal *et al.* [26]	**0.96513**	0.36109	**0.88916**
Drago *et al.* [12]	0.87215	0.17862	0.82714

Table 2. Quantitative measures for Figure 9.

Tone mapping operators	Structural fidelity	Naturalness	TMQI
Linear mapping	0.54176	4.7725×10^{-5}	0.66492
Manual-tuned Glozman *et al.*'s operator [22,24]	0.88690	**0.97412**	**0.96758**
Proposed automatic tone mapping [22,24]	0.89095	0.96408	0.96722
Meylan *et al.* [21]	0.90890	0.62874	0.92130
Reinhard *et al.* [18]	0.95811	0.83939	0.96642
Fattal *et al.* [26]	**0.97541**	0.36120	0.89174
Drago *et al.* [12]	0.95112	0.61756	0.93033

The k estimation model achieve our goal of deriving a simple automatic equation that can be used to estimate the value of k needed to produce tone-mapped images that have a good level of naturalness and structural fidelity. An advantage of the simplified automatic operator is that it depends solely on the estimated mean of the WDR image and without the additional computational complexity of computing the regional means of the standard deviation for each image, thereby making it easier to implement in hardware as part of the tone mapping architecture. The derived equation for estimating factor k is also an exponent-based function, which is easy to design and implement in hardware. Furthermore, because the image from the initial compressed stage is not needed for the actual tonal compression stage, this approach will not result in the addition of extra hardware memory for storing the full image from the estimation stage.

5. Proposed Hardware Tone Mapping Design

For the hardware implementation, a fixed-point precision instead of a floating point precision was used in the design, so as to reduce the hardware complexity, while still maintaining a good degree of accuracy when compared to a software implementation of the same system. The fixed-point representation chosen was 20 integer bits and 12 fractional bits. As a result, a total word length of 32 bits was used for the tone mapping system. The quantization error of $2.44140625 \times 10^{-4}$ is calculated for 12 fractional bits.

The actual tone mapping process is done on a pixel-by-pixel basis. First, a memory buffer is used to temporarily store the image; then, each pixel is continuously processed by the tone mapping system using modules that will be explained. The overview dataflow of the proposed tone mapping algorithm is described in Figure 10. The proposed tone mapping system contains five main modules: control unit, mean/max module, convolution module, factor k generator and inverse exponential module.

Figure 10. Dataflow chart for the hardware implementation of Glozman *et al.* [22,24].

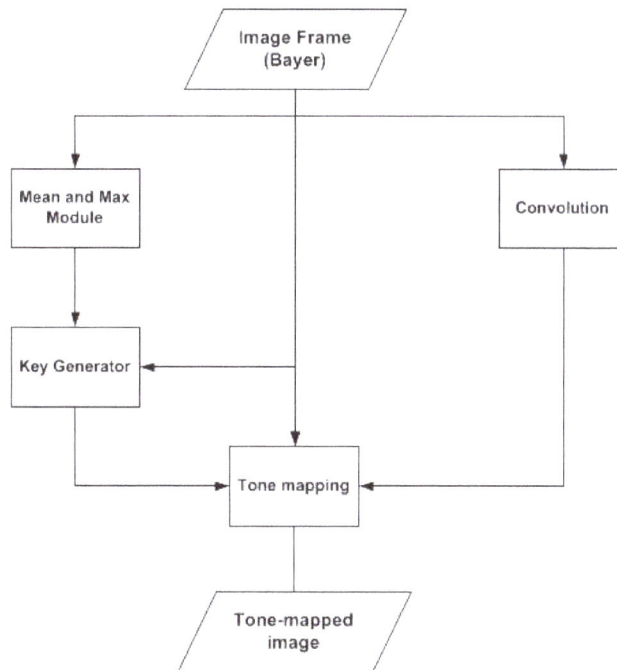

5.1. Convolution

In this hardware design, a 3×3 convolution kernel is used for obtaining the information of the neighboring pixels (Figure 11). This simple kernel is used so that the output pixels can be calculated without the use of multipliers. The convolution module is divided into two modules; the 3×3 data pixel input generation module and the actual smoothing module. To execute a pipelined convolution, two FIFO and three shift registers are used [47,48]. The structure of the convolution module is shown in Figure 12. The two FIFO and the three three-bit-shift registers act as line buffers for the convolution module. The FIFO is used to store a row of the image frame. At every clock cycle, a new input enters the pixel input generation module via the three-bit-shift register, and the line buffers get systematically updated resulting in new 3 × 3 input pixels (P1–P9). These updated nine input pixels (P1–P9) are then used by the smoothing module for convolution computation with the kernel (Figure 11).

Figure 11. Simple Gaussian smoothing kernel used for implementing convolution.

$$\frac{1}{16}\begin{bmatrix} 1 & 2 & 1 \\ 2 & 4 & 2 \\ 1 & 2 & 1 \end{bmatrix}$$

Figure 12. Convolution diagram for the field programmable gate array (FPGA).

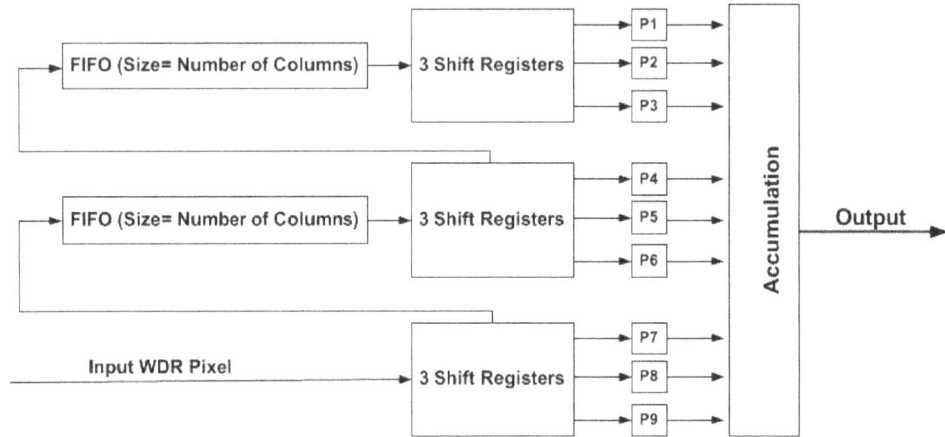

Inside the smoothing module, the output pixel is calculated by the use of simple adders and shift registers, thus removing the need for multipliers and dividers, which are expensive in terms of hardware resources. The convolution module has an initial latency of [number of pixel columns (nCols) + 12] clock cycles, because the FIFO buffers need to be filled before the convolution process can begin. This will result in the input pixel being stored in additional memory (FIFO), with the number of words being equal to the delay of the convolution system. To avoid this, the WDR input stored in the line buffer in the convolution module is used along with the convolution output of the same pixel location, for the actual tone mapping process in order to compensate for the delay, as well as to ensure that a pixel by pixel processing is maintained. Figure 13 shows a visual description of the readout system. The implemented approach is favorable for system-on-a-chip design, where the output from the CMOS image sensor can only be readout once and sent into the tone mapping architecture, thus saving memory, silicon area and power consumption.

Figure 13. (**a**) The implemented method of convolution and tone mapping operation, so as to reduce memory requirements; (**b**) the alternative method for the tone mapping architecture.

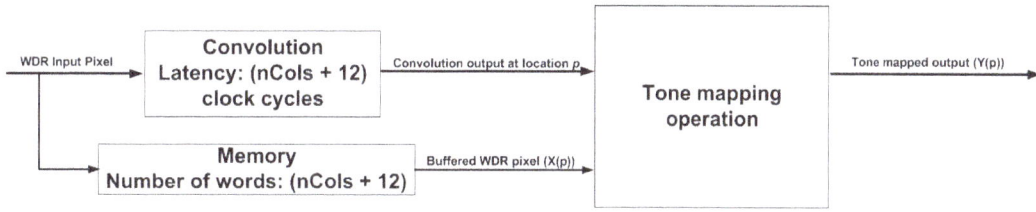

(**a**)

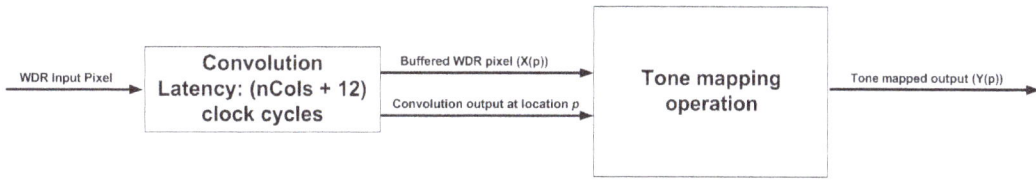

(**b**)

5.2. Factor k Generator

There are three regions (A,B,C) in all of the proposed method for estimating the value of k needed for tone mapping a specific WDR image [Equation (7)]. Region A and B are both constants that are represented in fixed-point in the hardware design. Since region B of the k-equation is a power function with a base of two and a decimal number as the exponent, a hardware approximation was implemented in order to achieve good accuracy for the computation of factor k in that region. The hardware modification is shown in Equation (8):

$$y = a \cdot 2^{bx} = a \cdot 2^{(I+F)} \tag{8}$$

where I and F are the integer and fractional parts of the value, bx. The multiplication by 2^I can be implemented using simple shift registers, so as to reduce hardware complexity. Since 2^F can be expressed in an exponential form as $e^{(\ln 2)F}$, the k-equation for region B becomes:

$$y = a \cdot 2^{(I+F)} = a \cdot 2^I \cdot e^{(\ln 2)F} \tag{9}$$

An iterative digit-by-digit exponent operation for computing exponential function $e^{(\ln 2)F}$ in fixed point precision was used. A detailed explanation of this operation can be found in [35,49]. It utilizes only shift registers and adders in order to reduce the amount of hardware resources needed for implementing exponential functions. A summary of the process is described below:

For any value of $x \in [0, \ln 2)$, $y = e^{(x)}$ can be approximated by setting up a data set of (x_i, y_i), where the initial values are $x_0 = x$ and $y_0 = 1$. Both x_i and y_i are calculated using the equations displayed below:

$$y_i = e^{-x_i} \cdot e^x \tag{10}$$

$$x_{i+1} = x_i - \ln(b_i) \tag{11}$$

where b_i is a constant that is equal to $1 + s_i \cdot 2^{-i}$, and i ranges from zero to 11. s_i is equal to zero or one, depending on the conditions shown in Equation (12).

$$s = \begin{cases} 1, & \text{if } x_i \geq \ln(1 + 2^{-i}) \\ 0, & \text{otherwise} \end{cases} \qquad (12)$$

Since we are implementing 12 fractional bits for the system's fixed-point precision, a 12×12-bit read-only memory (ROM) is used as the lookup table for obtaining $\ln(b_i)$. x_i and y_i are computed repetitively using the conditions displayed in Equations (13) and (14).

$$y_{i+1} = \begin{cases} (1 + 2^{-i}) \cdot y_i, & \text{if } s = 1 \\ y_i, & \text{otherwise} \end{cases} \qquad (13)$$

$$x_{i+1} = \begin{cases} x_i - \ln(1 + 2^{-i}), & \text{if } s = 1 \\ x_i, & \text{otherwise} \end{cases} \qquad (14)$$

A shift and add operation can be used in computing y_{i+1} in Equation (13) when $s = 1$. Notice that in Equation (10), when $x_i = 0$, then $y_i = e^x$, which is the final exponential value, since in our case, it is $e^{(\ln 2)F}$, where $0 \leq F \ln 2 < \ln 2$. Taking $x_0 = F \cdot \ln 2$, we can use the iteration method explained above to calculate the exponential part of the k-equation [Equation (9)]. The k value obtained is stored and used for the next frame tone mapping operation, so that the processing time is not increased due to the addition of a k-generator.

5.3. Inverse Exponential Function

The inverse exponential function is important for the tone mapping algorithm by Glozman *et al.* [22,24], and it is described in Chapter 2. A modification of the digit-by-digit algorithm for implementing exponential functions on hardware was made [35]. It is similar to the exponential operation described above; however, in this case, because the tone mapping algorithm is an inverse exponential function, we will require hardware dividers if the original digit-by-digit algorithm was implemented. This will increase the hardware complexity and latency of the system. Therefore, the digit-by-digit algorithm was modified to suit this application. The main part of the original algorithm that is needed for this module and the modification of the algorithm are described below. For any value of x, it can be represented as shown in Equation (15):

$$x = (I + f) \cdot \ln 2 \qquad (15)$$

where I and f are the integer and fractional parts of the value, x. To obtain I and f, the equation is rearranged as described in Equation (16).

$$I + f = x \cdot \log_2 e \qquad (16)$$

Therefore, for any arbitrary value of $y = \exp x$, it can be represented using Equation (17):

$$y = e^x = e^{(I+f) \cdot \ln 2} = e^{I \cdot \ln 2 + f \cdot \ln 2} = 2^I e^{f \cdot ln2} \qquad (17)$$

Shift operation is also used to implement 2^I in order to reduce the hardware complexity of the module. Since f is a fractional number that lies between zero and one, $f \cdot \ln 2$ therefore ranges from zero to $\ln 2$. Then, the Taylor series approximation (18) of the exponential equation (until the third order) can be used to calculate $e^{f \cdot \ln 2}$.

$$e^{f \cdot \ln 2} = 1 + \frac{f \cdot \ln 2}{1!} + \frac{(f \cdot \ln 2)^2}{2!} + \frac{(f \cdot \ln 2)^3}{3!} \tag{18}$$

Since the actual equation used in the tone mapping algorithm is actually in an inverse exponential-based form, the following assumptions were made, as shown in Equation (19).

$$y = 1 - e^{-x} = \begin{cases} A, & \text{if } x \leq 8 \\ 1, & \text{if } x > 8 \end{cases} \tag{19}$$

where $A = 1 - 2^{-I} \cdot e^{-f \cdot \ln 2}$. The assumption in Equation (19) was made because $e^{-9} = 0.00012341$, which is less than the quantization error of using 12 fractional bits for the tone mapping architecture (0.00024414). Using the Taylor series approximation of the equation $e^{-f \cdot \ln 2}$ and the conditions in Equation (19), the overall equation is displayed below:

$$y = 1 - e^{-x} = \begin{cases} 1 - \left[2^{-I} \times \left(1 - \frac{f \cdot \ln 2}{1!} + \frac{(f \cdot \ln 2)^2}{2!} - \frac{(f \cdot \ln 2)^3}{3!} \right) \right], & \text{if } x \leq 8 \\ 1, & \text{if } x > 8 \end{cases} \tag{20}$$

The hardware implementation of the inverse exponential function was performed in a pipelined manner, so as to facilitate the per pixel processing of the WDR image. The latency of the module is seven clock cycles.

5.4. Control Unit

The control unit of the tone mapping system is implemented as finite state machines (FSM). It is used to generate the control signals to synchronize the overall tone mapping process (Figure 14). The operations and data flow of the whole tone mapping system is driven by the control unit.

The hardware implementation has three main states:

- **S0**: In this state, all modules remain inactive until the enable signal, **Start**, is set as high and the FSM moves to state **S1**;

- **S1**: In this state, the actual tone mapping operation is performed. An enable signal is sent to the external memory buffer, which results in one input WDR pixel being read into the system at each clock cycle. This input pixel is shared by the mean, maximum, factor k estimation block and image convolution. The input pixels from the frame buffer are used in computing the mean, maximum, factor k estimation block and image convolution. The filtered output values from the convolution module are used along with the input pixels for the actual computation of the tone mapping algorithm and the estimation of factor k. The mean, maximum and factor k of the previous frame are used in current frame tone mapping computation. Once all four computations have been completed and stored, an enable signal, **endtone**, is set high, and the FSM moves to state **S2**;

- **S2**: In this state, the mean, maximum and factor k of the current WDR frame are stored for the next frame image processing. An end signal, **finished**, is set high, and the FSM returns to state **S0**. The whole operation is repeated if there is another WDR frame to be processed.

Figure 14. State diagram of the controller for the tone mapping system.

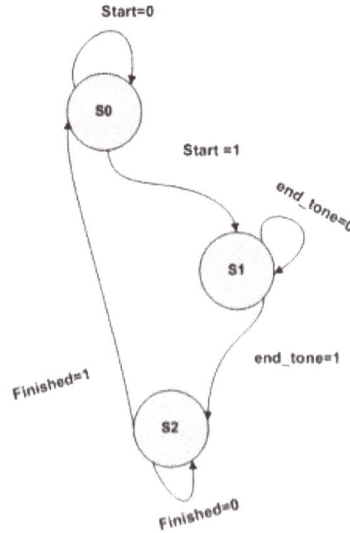

In order to implement the system efficiently, all of the modules were executed in a pipelined and parallel manner. In total, four dividers were used in the design in which three dividers were 44 bits long and one divider was a 14-bit word length for the fixed point division. A more descriptive image of the processing stages involved in the hardware design implemented is shown in Figures 15 and 16.

Figure 15. Diagram of the pathway of the input in the hardware implementation of the tone mapping system.

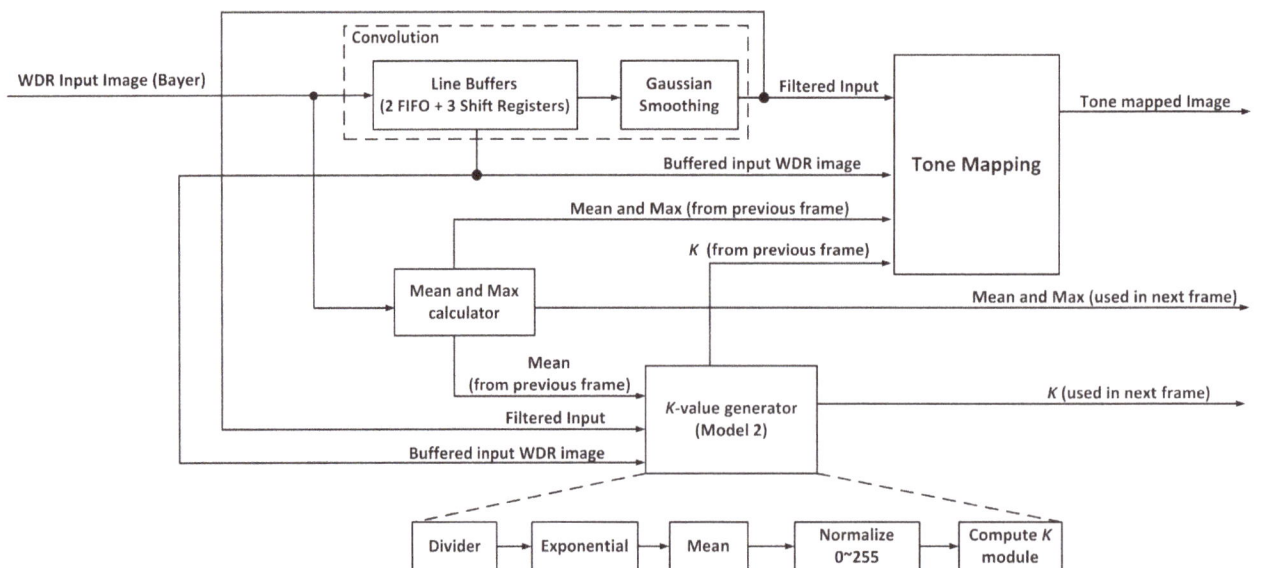

Figure 16. Detailed diagram of the pathway for the tone mapping operation in hardware.

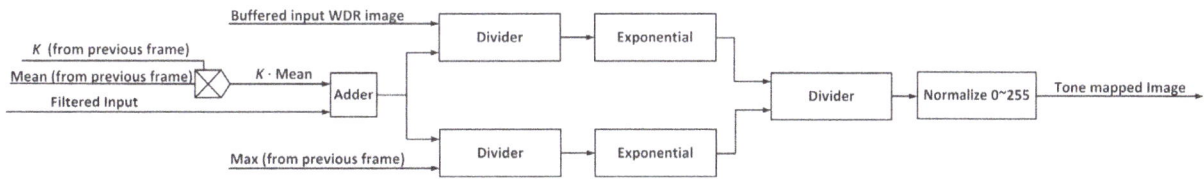

It should be noted that the mean and maximum values of the current frame are stored for processing the next frame, while the mean and maximum values of the previous frame are used in the tone mapping operation of the present frame. Since the k computation also requires the mean of the compressed image frame, the k value of a previous frame can be used. This is because when operating in real-time, such as 60 fps, there are very little variations in information from one frame to the next [50]. This saves in terms of processing speed and removes the need to store the full image memory in order to calculate the mean and maximum and, then, implement the tone mapping algorithm. As a result, power consumption will also be reduced in comparison to a system that would have required the use of an external memory buffer for storing the image frame.

6. Experimental Results

WDR images from the Debevec library and other sources were used to test the visual quality of both hardware architectures. These WDR images were first converted to the Bayer CFA pattern on MATLAB®. Then, the Bayer result was converted into fixed-point representation and used as an input into the hardware implementation of the tone mapping operator. The tone mapped output image was converted back to the RGB image format using a simple bilinear interpolation on MATLAB®. Output images from functional simulation using ModelSim® for WDR images are shown in Figures 17–19.

Figure 17. Nave image [image size (1024 × 768)] [51]. (**a**) Simulation results; (**b**) Hardware results.

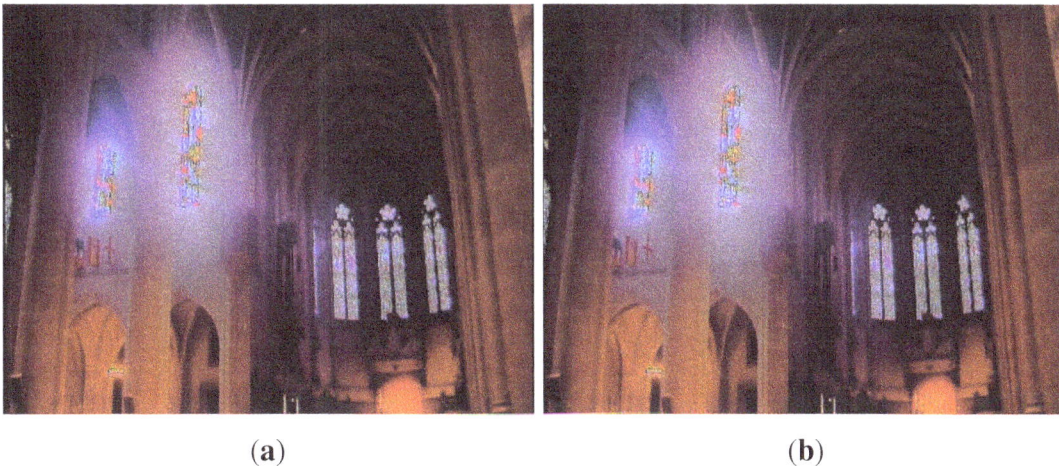

(**a**)   (**b**)

Figure 18. Bird of paradise image [image size (1024 × 768)] [29]. (**a**) Simulation results; (**b**) Hardware results.

(**a**)   (**b**)

Figure 19. Raw image [image size (1152 × 768)]. (**a**) Simulation results; (**b**) Hardware results.

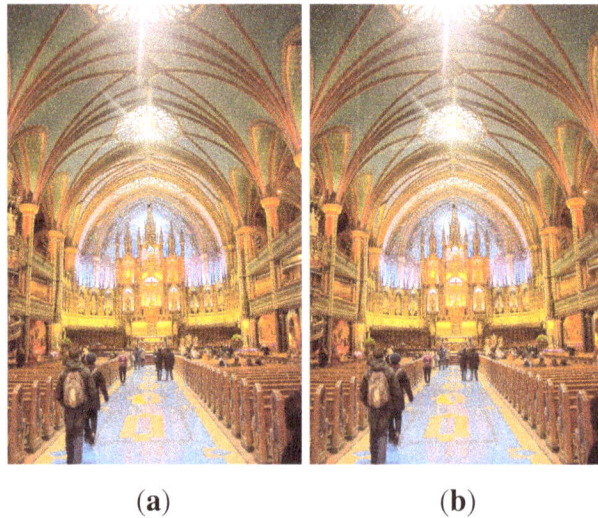

(**a**)   (**b**)

6.1. Image Quality: Peak Signal-to-Noise Ratio (PSNR)

The peak signal-to-noise ratio (PSNR) of the results from the hardware simulation on ModelSim® was calculated in order to quantify the amount of error produced due to the use of fixed-point precision. The results from the floating point implementation on MATLAB® were used in this quantitative evaluation. Using Equations (21) and (22), the PSNR was calculated:

$$PSNR = 10 \log_{10} \left[\frac{255^2}{MSE} \right] \tag{21}$$

$$MSE = \frac{1}{MN} \sum_{m=0}^{M-1} \sum_{n=0}^{N-1} \left[f(m,n) - k(m,n) \right]^2 \tag{22}$$

where MSE is the mean-squared error between the output images obtained from the floating point implementation and the fixed-point implementation on hardware. Higher PSNR values indicate better visual image similarity between the two images being compared. The PSNR values of implementations are shown in Table 3. It should be noted that the hardware-friendly approximations made while implementing the k generator module may have affected the PSNR of the tone-mapped images. Even with that, the results show that the hardware implementation of the tone mapping algorithm produce acceptable results that can be compared to the MATLAB® simulations.

Table 3. peak signal-to-noise ratio (PSNR) of images tested.

Image (N × M)	Implementation PSNR (dB)
Nave (720 × 480)	56.11
Desk (874 × 644)	46.21
Bird of Paradise (1024 × 768)	49.63
Memorial (512 × 768)	54.27
Raw Image (1152 × 768)	54.85

6.2. Image Quality: Structural Similarity (SSIM)

SSIM is an objective image quality assessment tool that is used for measuring the similarity between two images [52]. The SSIM index obtained is used to evaluate how one image is structurally similar to another image that is regarded as perfect quality. Using Equation (23), the SSIM was calculated [52]:

$$SSIM = \frac{(2\mu_x\mu_y + C_1)(2\sigma_{xy} + C_2)}{(\mu_x{}^2 + \mu_y{}^2 + C_1)(\sigma_x{}^2 + \sigma_y{}^2 + C_2)} \tag{23}$$

where x and y represent two image patches obtained from the same spatial location in the two images being compared. The mean of x and the mean of y are represented by μ_x and μ_y, respectively. The covariance of x and y is represented by σ_{xy}. The variance of x and the variance of y are denoted by σ_x^2 and σ_y^2, respectively. The variances are used to interpret the contrast of the image, while the mean is used to represent the luminance of the image [52]. The constants, C_1 and C_2, are used to prevent instability in the division. They are calculated based on Equations (24) and (25):

$$C_1 = (K_1 \cdot L)^2 \tag{24}$$

$$C_2 = (K_2 \cdot L)^2 \tag{25}$$

where L is the dynamic range of the pixel values (255- for eight-bit tone-mapped images). K_1 and K_2 are 0.01 and 0.03, respectively. The closer the SSIM values are to one, the higher the structural similarity between the two images being compared. The SSIM values of images are shown in Table 4.

Table 4. Structural similarity (SSIM) of the images tested.

Image (N × M)	Implementation SSIM
Nave (720 × 480)	0.9999
Desk (874 × 644)	0.9998
Bird of Paradise (1024 × 768)	0.9998
Memorial (512 × 768)	0.9998
Raw image (1152 × 768)	0.9997

6.3. Performance Analysis: Hardware Cost, Power Consumption and Processing Speed

The FPGA implementation of the overall system was evaluated in terms of the hardware cost, power consumption and processing speed. In Table 5, a summary of the synthesis report from Quartus v11.1 for the Stratix II and Cyclone III devices is shown for the architecture. The system was designed in such a way that the total number of logic cells and registers used would not dramatically increase if the size of the image became greater than 1024 × 768. However, the number of memory bits would increase because of the two-line buffers used in the convolution module.

In addition, since power consumption is of importance for possible applications, such as mobile devices and still/video cameras, ensuring that the system consumes low power is of great importance. Using the built in PowerPlay power analyzer tool of Altera® Quartus v11.1, the power consumption of the full tone mapping system was estimated for the Cyclone III device. The power consumed by the optimized tone mapping system when operating at a clock frequency of 50 MHz was estimated to be 250 mW.

Table 5. FPGA synthesis report for a 1024 × 768 multi-fused WDR image.

Stratix II EP2S130F780C4	Used	Available	Percentage (%)
Combinatorial ALUTs	8,546	106,032	8.06
Total registers	10,442	106,032	9.85
DSP block nine-bit elements	60	504	11.90
Memory bits	68,046	6,747,840	1.00
Maximum frequency (MHz)	114.90	N/A	N/A
Operating frequency (MHz)	50	N/A	N/A
Cyclone III EP3C120F48417	**Used**	**Available**	**Percentage (%)**
Combinatorial functions	12,154	119,088	10.21
Total registers	10,518	119,088	8.83
Embedded multipliers nine-bit elements	36	576	6.25
Memory bits	68,046	3,981,312	1.71
Maximum frequency (MHz)	116.50	N/A	N/A
Operating frequency (MHz)	50	N/A	N/A

For the hardware architecture, the maximum frequency obtained for the Stratix II and Cyclone III were 114.90 MHz and 116.50 MHz, respectively. In order to calculate the processing time for one N × M pixel resolution, Equation (26) is used.

$$T = [\text{Total clock cycles per frame}] \cdot \frac{1}{\text{operating clock frequency}} \tag{26}$$

$$\text{Total clock cycles per frame} = M \cdot N + N + 100 \tag{27}$$

This implies that for a 1024 × 768 image using this hardware implementation, the processing time for tone mapping of each frame is 7.9 ms (126 fps) and 3.45 ms (63 fps) at a clock frequency of 100 MHz and 50 MHz, respectively. This means that the system can work in real-time if the input video rate is 60 fps at a lower clock frequency of 50 MHz.

6.4. Comparison with Existing Research

In Table 6, a comparison of the proposed tone mapping architecture with other tone mapping hardware systems in terms of processing speed, logic elements, memory, power consumption and PSNR is shown. All the FPGA-based (Altera®) systems compared utilized a Stratix II device. Due to proprietary reasons, there is no way of converting the hardware resources requirements for the Xilinx®-based tone mapping system by Ureña et al. [25,37] to an Altera® system. In this case, only properties, such as memory requirements, power consumption and processing speed, are compared with the other hardware-based systems.

The proposed full tone mapping system utilizes less hardware resources (logic elements and memory) in comparison to the other hardware tone mapping systems published. This is because of the low complexity of implementing the automatic rendering algorithm and tone mapping algorithm, since they are both exponent-based functions. In addition, proposed hardware architecture achieves higher and similar processing times in comparison to the other hardware-based systems implemented. Furthermore, the power consumption estimated for the proposed tone mapping algorithm using the Quartus power analyzer was lower in contrast to the only found power results for a tone mapping system implemented on an FPGA. Ureña et al.'s hardware architecture consumed 4× more power in contrast to our systems that implemented a larger image size (1024 × 768). Taking into account that if we implement our full tone mapping algorithm in ASIC rather than FPGA [53], our power consumption is anticipated to decrease substantially. Thus, we can conclude that our system achieves our goals of being low power-efficient.

It should be noted that among the hardware systems compared, only this current hardware architecture implements a rendering parameter selection unit that automatically adjusts the rendering parameter's value based on the properties of the image being processed. Hassan et al. [33], Carletta et al. [34] and Vylta et al. [36] selected a constant value for their rendering parameters, while Chiu et al. [35] and Ureña et al. [25] required the manual adjustment of their rendering parameters. This is not favorable for a real-time video processing applications, where the parameters need to be constantly adjusted or a constant value may not work for varying image key.

Table 6. Comparison with other tone mapping hardware implementation. (fps: frames per second; SOC: system-on-a-chip).

Research Works	Tone mapping algorithm	Type of tone mapping operator	Design Type	Frame size	Image type	Maximum operating frequency (MHz)	Processing speed (fps)	Logic elements	Memory (bits)	Power (mW)	PSNR (dB) (memorial image)
This work	Improved Glozman et al. [22,24]	Global and local	FPGA (Altera)	1,024 × 768	Color	114.90	126	8,546	68,046	250	54.27
Hassan et al. [33]	Reinhard et al. [18]	Local	FPGA (Altera)	1,024 × 768	Monochrome	77.15	60	34,806	3,153,408	N/A	33.70
Carletta et al. [34]	Reinhard et al. [18]	Local	FPGA (Altera)	1,024 × 768	Color	65.86	N/A	44,572	3,586,304	N/A	N/A
Vylta et al. [36]	Fattal et al. [26]	Local	FPGA (Altera)	1 Megapixel	Color	114.18	100	9,019	307,200	N/A	53.83
Chiu et al.[35]	Reinhard et al. [18]/ Fattal et al. [26]	Global / Local	ARM-SOC	1,024 × 768	N/A	100	60	N/A	N/A	177.15	45/35
Ureña et al. [25,37]	Ureña et al. [25,37]	Global and local	FPGA (Xilinx)	640 × 480	Color	40.25	60	N/A	718,848	900	N/A

7. Conclusions

In this paper, we proposed an algorithm that can be used in estimating the rendering parameter k, needed for Glozman *et al.*'s operator. The automatic rendering algorithm requires the image statistics of the WDR image in order to determine the value of the rendering parameter k that will be used. An approximation of the independent property of the WDR image is obtained by initially compressing the WDR image in order to have a standard platform for obtaining the rendering parameter for different WDR images. This compressed image is not used for the actual tone mapping process by Glozman *et al.*'s operator, and it is mainly for configuration purposes. Hence, no additional memory is needed for storing the initial compressed image frame. The proposed algorithm was able to estimate the rendering parameter k that was needed for tone mapping a variety of WDR images.

A hardware-based implementation of the automatic rendering algorithm along with Glozman *et al.*'s tone mapping algorithm was also presented in this paper. The hardware architecture produced output images with good visual quality and high PSNR and SSIM values in comparison to the compressed images obtained from the software simulation of the tone mapping system. The hardware implementation produced output images in Bayer CFA format that can be converted to an RGB color image by using simple demosaicing techniques on hardware. Low hardware resources and power consumption were also achieved in comparison to other hardware-based tone mapping systems published. WDR images were processed in real-time, because of approximations made when designing the hardware architecture. Potential applications, such as medical imaging, mobile phones, video surveillance systems and other WDR video applications, will benefit from a real-time embedded device with low hardware resources and automatic configuration.

The overall architecture is well compacted, and it does not require the constant tuning of the rendering parameters for each WDR image frame. Additionally, it requires low hardware resources and no external memory for storing the full image frame, thereby making it adequate as a part of a system-on-a-chip. Future research may concentrate on embedding the tone mapping system as a part of a system-on-a-chip with WDR imaging capabilities. In order to achieve this, the WDR image sensor should also have an analog-to-digital converter (ADC) as part of the system. In addition, characteristics of the automatic generation of the rendering parameter k with respect to flickering artifacts during WDR video compression should be investigated in future works.

Conflicts of Interest

The authors declare no conflict of interest.

References

1. Yadid-Pecht, O. Wide-dynamic-range sensors. *Opt. Eng.* **1999**, *38*, 1650–1660.
2. Fish, A.; Yadid-Pecht, O. APS Design: From Pixels to Systems. In *CMOS Imagers: From Phototransduction to Image Processing*; Springer: New York, NY, USA, 2004; pp. 102–139.

3. Debevec, P.E.; Malik, J. Recovering High Dynamic Range Radiance Maps from Photographs. In Proceedings of the 24th Annual Conference on Computer Graphics and Interactive Techniques, Los Angeles, CA, USA, 3–8 August 1997; pp. 369–378.

4. Yadid-Pecht, O.; Belenky, A. Autoscaling CMOS APS with Customized Increase of Dynamic Range. In Proceedings of IEEE International Solid Sate Circuits Conference (ISSCC), San Francisco, CA, USA, 7 February 2001; pp. 100–101.

5. Yang, D.; Gamal, A.; Fowler, B.; Tian, H. A 640 × 512 CMOS image sensor with ultrawide dynamic range floating-point pixel-level ADC. *IEEE J. Solid-State Circuits* **1999**, *34*, 1821–1834.

6. Dattner, Y.; Yadid-Pecht, O. High and low light CMOS imager employing wide dynamic range expansion and low noise readout. *IEEE Sens. J.* **2012**, *12*, 2172–2179.

7. Spivak, A.; Teman, A.; Belenky, A.; Yadid-Pecht, O.; Fish, A. Low-voltage 96 dB snapshot CMOS image sensor with 4.5 nW power dissipation per pixel. *Sensors* **2012**, *12*, 10067–10085.

8. Schanz, M.; Nitta, C.; Bussmann, A.; Hosticka, B.; Wertheimer, R. A high dynamic range CMOS image sensor for automotive applications. *IEEE J. Solid-State Circuits* **1999**, *35*, 932–938.

9. Devlin, K. *A Review of Tone Reproduction Techniques*; Technical Report CSTR-02-005; Department of Computer Science, University of Bristol: Bristol, UK, 2002.

10. Brainard, D.H.; Wandell, B.A. Analysis of the retinex theory of color vision. *J. Opt. Soc. Am. A* **1986**, *3*, 1651–1661.

11. Tumblin, J.; Rushmeier, H. Tone reproduction for realistic images. *IEEE Comput. Graph. Appl.* **1993**, *13*, 42–48.

12. Drago, F.; Myszkowski, K.; Annen, T.; Chiba, N. Adaptive logarithmic mapping for displaying high contrast scenes. *Comput. Graph. Forum* **2003**, *22*, 419–426.

13. Qiu, G.; Duan, J. An Optimal Tone Reproduction Curve Operator for the Display of High Dynamic Range Images. In Proceedings of the IEEE International Symposium on Circuits and Systems (ISCAS), Kobe, Japan, 23–26 May 2005; Volume 6, pp. 6276–6279.

14. Liu, C.H.; Au, O.; Wong, P.; Kung, M. Image Characteristic Oriented Tone Mapping for High Dynamic Range Images. In Proceedings of the IEEE International Conference on Multimedia and Expo, Hannover, Germany, 23–26 June 2008; pp. 1133–1136.

15. Pattanaik, S.N.; Tumblin, J.; Yee, H.; Greenberg, D.P. Time-Dependent Visual Adaptation for Fast Realistic Image Display. In Proceedings of the 27th Annual Conference on Computer Graphics and Interactive Techniques, New Orleans, LA, USA, 23–28 July 2000; pp. 47–54.

16. Ashikhmin, M. A Tone Mapping Algorithm for High Contrast Images. In Proceedings of the 13th Eurographics workshop on Rendering, Pisa, Italy, 26–28 June 2002; pp. 145–156.

17. Kuang, J.; Johnson, G.M.; Fairchild, M.D. iCAM06: A refined image appearance model for HDR image rendering. *J. Vis. Commun. Image Represent.* **2007**, *18*, 406–414.

18. Reinhard, E.; Stark, M.; Shirley, P.; Ferwerda, J. Photographic tone reproduction for digital images. *ACM Trans. Graph.* **2002**, *21*, 267–276.

19. Rahman, Z.; Jobson, D.; Woodell, G. *A Multiscale Retinex for Color Rendition and Dynamic Range Compression*; Technical Report for NASA: Langley, VA, USA, 1996.

20. Herscovitz, M.; Yadid-Pecht, O. A modified multiscale retinex algorithm with an improved global impression of brightness for wide dynamic range pictures. *Mach. Vis. Appl.* **2004**, *15*, 220–228.

21. Meylan, L.; Alleysson, D.; Süsstrunk, S. Model of retinal local adaptation for the tone mapping of color filter array images. *J. Opt. Soc. Am. A* **2007**, *24*, 2807–2816.

22. Ofili, C.; Glozman, S.; Yadid-Pecht, O. An in-depth analysis and image quality assessment of an exponent-based tone mapping algorithm. *Inf. Models Anal.* **2012**, *1*, 236–250.

23. Meylan, L.; Susstrunk, S. High dynamic range image rendering with a retinex-based adaptive filter. *IEEE Trans. Image Process.* **2006**, *15*, 2820–2830.

24. Glozman, S.; Kats, T.; Yadid-Pecht, O. Exponent operator based tone mapping algorithm for color wide dynamic range. *IEEE Trans. Image Process.*, 2011, Submitted for publication.

25. Ureña, R.; Martínez-Cañada, P.; Gómez-López, J.M.; Morillas, C.A.; Pelayo, F.J. Real-time tone mapping on GPU and FPGA. *EURASIP J. Image Video Process.* **2012**, *2012*, doi:10.1186/1687-5281-2012-1.

26. Fattal, R.; Lischinski, D.; Werman, M. Gradient domain high dynamic range compression. *ACM Trans. Graph.* **2002**, *21*, 249–256.

27. Artyomov, E.; Fish, A.; Yadid-Pecht, O. Image sensors in security and medical applications. *Inf. Theor. Appl.* **2007**, *14*, 115–127.

28. Kats, T.; Glozman, S.; Yadid-Pecht, O. Efficient color filter array luminance LOG based algorithm for wide dynamic range (WDR) images compression. *Opt. Eng.*, 2011, Submitted for publication.

29. Reinhard, E.; Ward, G.; Pattanaik, S.; Debevec, P. *High Dynamic Range Imaging: Acquisition, Display, and Image-Based Lighting (The Morgan Kaufmann Series in Computer Graphics)*; Morgan Kaufmann Publishers Inc.: San Francisco, CA, USA, 2005.

30. Oppenheim, A.; Schafer, R.; Stockham, T.G., Jr. Nonlinear filtering of multiplied and convolved signals. *IEEE Proc.* **1968**, *56*, 1264–1291.

31. Durand, F.; Dorsey, J. Fast Bilateral Filtering for the Display of High-Dynamic-Range Images. In Proceedings of the 29th Annual Conference on Computer Graphics and Interactive Techniques, San Antonio, TX, USA, 21–26 July 2002; pp. 257–266.

32. Lee, C.; Kim, C. Gradient Domain Tone Mapping of High Dynamic Range Videos. In Proceedings of the IEEE International Conference on Image Processing (ICIP), San Antonio, TX, USA, 16–19 September 2007; Volume 3, pp. III:461–III:464.

33. Hassan, F.; Carletta, J. An FPGA-based architecture for a local tone-mapping operator. *J. Real-Time Image Process.* **2007**, *2*, 293–308.

34. Carletta, J.E.; Hassan, F.H. Method for Real-Time Implementable Local Tone Mapping for High Dynamic Range Images. US Patent 20090041376, 19 February 2009.

35. Chiu, C.T.; Wang, T.H.; Ke, W.M.; Chuang, C.Y.; Huang, J.S.; Wong, W.S.; Tsay, R.S.; Wu, C.J. Real-time tone-mapping processor with integrated photographic and gradient compression using 0.13 µm technology on an Arm Soc platform. *J. Signal Process. Syst.* **2011**, *64*, 93–107.

36. Vytla, L.; Hassan, F.; Carletta, J. A real-time implementation of gradient domain high dynamic range compression using a local poisson solver. *J. Real-Time Image Process.* **2013**, *8*, 153–167.

37. Martínez-Cañada, P.; Morillas, C.A.; Ureña, R.; Gómez-López, J.M.; Pelayo, F.J. Embedded system for contrast enhancement in low-vision. *J. Syst. Archit.* **2013**, *59*, 30–38.

38. Kiser, C.; Reinhard, E.; Tocci, M.; Tocci, N. Real Time Automated Tone Mapping System for HDR Video. In Proceedings of the IEEE International Conference on Image Processing, Orlando, FL, USA, 30 September–3 October 2013.

39. Reinhard, E. Parameter estimation for photographic tone reproduction. *J. Graph. Tools* **2002**, *7*, 45–52.

40. Tamburrino, D.; Alleysson, D.; Meylan, L.; Süsstrunk, S. Digital Camera Workflow for High Dynamic Range Images Using a Model of Retinal Processing. In Proceedings of the IS&T/SPIE Electronic Imaging: Digital Photography IV, San Jose, CA, USA, 28–29 January 2008; Volume 6817.

41. Jobson, D.J.; Rahman, Z.; Woodell, G.A. The statistics of visual representation. *Proc. SPIE* **2002**, *4736*, 25–35.

42. Yeganeh, H.; Wang, Z. Objective quality assessment of tone-mapped images. *IEEE Trans. Image Process.* **2013**, *22*, 657–667.

43. Cadik, M.; Slavik, P. The Naturalness of Reproduced High Dynamic Range Images. In Proceedings of the 9th International Conference on Information Visualisation, London, UK, 6–8 July 2005; pp. 920–925.

44. Watkins, A.; Scheaffer, R.; Cobb, G. *Statistics: From Data to Decision*; Wiley: Hoboken, NJ, USA, 2010.

45. MathWorks. *Curve Fitting Toolbox 1: User's Guide*; MathWorks: Natick, MA, USA, 2006.

46. Luminance HDR. Available online: http://qtpfsgui.sourceforge.net/ (accessed on 7 November 2011).

47. Guo, Z.; Xu, W.; Chai, Z. Image Edge Detection Based on FPGA. In Proceedings of the 9th International Symposium on Distributed Computing and Applications to Business Engineering and Science (DCABES), Hong Kong, China, 10–12 August 2010; pp. 169–171.

48. Benedetti, A.; Prati, A.; Scarabottolo, N. Image Convolution on FPGAs: The Implementation of a Multi-FPGA FIFO Structure. In Proceedings of the 24th Euromicro Conference, Vasteras, Sweden, 27 August 1998; Volume 1, pp. 123–130.

49. Kantabutra, V. On hardware for computing exponential and trigonometric functions. *IEEE Trans. Comput.* **1996**, *45*, 328–339.

50. Iakovidou, C.; Vonikakis, V.; Andreadis, I. FPGA implementation of a real-time biologically inspired image enhancement algorithm. *J. Real-Time Image Process.* **2008**, *3*, 269–287.

51. Ward, G. High Dynamic Range Image Examples. Available online: http://www.anyhere.com/gward/hdrenc/pages/originals.html (accessed on 7 November 2011).

52. Wang, Z.; Bovik, A.; Sheikh, H.; Simoncelli, E. Image quality assessment: from error visibility to structural similarity. *IEEE Trans. Image Process.* **2004**, *13*, 600–612.

53. Kuon, I.; Rose, J. Measuring the gap between FPGAs and ASICs. *IEEE Trans. Comput.-Aided Des. Integr. Circuits Syst.* **2007**, *26*, 203–215.

7

Three-Dimensional Wafer Stacking Using Cu TSV Integrated with 45 nm High Performance SOI-CMOS Embedded DRAM Technology [†]

Pooja Batra [1,*], Spyridon Skordas [2], Douglas LaTulipe [3], Kevin Winstel [2], Chandrasekharan Kothandaraman [1], Ben Himmel [3], Gary Maier [1], Bishan He [1], Deepal Wehella Gamage [1], John Golz [1], Wei Lin [2], Tuan Vo [3], Deepika Priyadarshini [2], Alex Hubbard [2], Kristian Cauffman [2], Brown Peethala [2], John Barth [3], Toshiaki Kirihata [1], Troy Graves-Abe [1], Norman Robson [1] and Subramanian Iyer [1]

[1] IBM Corporation Systems and Technology Group, Hopewell Junction, NY 12533, USA;
E-Mails: raman1@us.ibm.com (C.K.); gmaier@us.ibm.com (G.M.); dwehella@us.ibm.com (B.H.); bishanhe@us.ibm.com (D.W.G.); golzj@us.ibm.com (J.G.); nrobson@us.ibm.com (T.K.); tlgraves@us.ibm.com (T.G.-A.); kirihata@us.ibm.com (N.R.); ssiyer@us.ibm.com (S.I.)

[2] IBM Corporation Systems and Technology Group, Albany, NY 12203, USA;
E-Mails: skordas@us.ibm.com (S.S.); winstel@us.ibm.com (K.W.); linw@us.ibm.com (W.L.); dpriyad@us.ibm.com (D.P.); arhubbar@us.ibm.com (A.H.); kpcauffm@us.ibm.com (K.C.); peethala@us.ibm.com (B.P.)

[3] Formerly with IBM Corporation Systems and Technology Group;
E-Mails: dlip410@gmail.com (D.L.); benhimmel@gmail.com (B.H.); tvo@albany.edu (T.V.); barth65@gmail.com (J.B.)

[†] Conference version published in IEEE-S3S, 2013.

[*] Author to whom correspondence should be addressed; E-Mail: prbatra@us.ibm.com

Abstract: For high-volume production of 3D-stacked chips with through-silicon-vias (TSVs), wafer-scale bonding offers lower production cost compared with bump bond technology and is promising for interconnect pitches smaller than 5 μ using available tooling. Prior work has presented wafer-scale integration with tungsten TSV for low-power applications.

This paper reports the first use of low-temperature oxide bonding and copper TSV to stack high performance cache cores manufactured in 45 nm Silicon On Insulator-Complementary Metal Oxide Semiconductor (SOI-CMOS) embedded DRAM (EDRAM) having 12 to 13 copper wiring levels per strata and upto 11000 TSVs at 13 μm pitch for power and signal delivery. The wafers are thinned to 13 μm using grind polish and etch. TSVs are defined post bonding and thinning using conventional alignment techniques. Up to four additional metal levels are formed post bonding and TSV definition. A key feature of this process is its compatibility with the existing high performance POWER7™ EDRAM core requiring neither modification of the existing CMOS fabrication process nor re-design since the TSV RC characteristic is similar to typical 100–200 μm length wiring load enabling 3D macro-to-macro signaling without additional buffering Hardware measurements show no significant impact on device drive and off-current. Functional test at wafer level confirms 2.1 GHz 3D stacked EDRAM operation.

Keywords: EDRAM; 3D; SOI; through-silicon-via (TSV); wafer stacking

1. Introduction

As semiconductor technology scaling becomes increasingly difficult, 3D stacking has recently gained attention due to its potential to offer much higher form factor along with higher performance and lower power compared to 2D designs. 3D stacking can be achieved in many different ways: package-to-package stack, die-to-die stack, die-to-wafer stack, and wafer-to-wafer stack. Package-to-package stack and die-to-die stack allow selection of known good dies for stacking and, thus, provide higher yield but limited performance improvement compared to 2D. In addition, these have higher cost as each die needs to be handled individually while building a stacked module [1–3]. The through-silicon-via (TSVs) used in package and die level stacks are typically larger in size thereby limiting the bandwidth and performance offered by these. On the other end, wafer-to-wafer 3D stack allows TSVs to be scaled by 20 times thereby allowing much higher bandwidth and performance improvement along with lowest manufacturing cost by allowing stacking of wafers instead of chip and thereby supporting volume production. However wafer-stacking technology can potentially suffer from compounded yield loss, but this can be mitigated by the use of innovative circuit design, fault tolerant and repair techniques. Due to the strong consumer demand for higher performance in a smaller size with lower cost 3D technology, the industry is moving towards wafer level stacked technology.

The scaling of TSVs depends on the aspect ratio (D_{TSV}/T_{Si}) of TSV diameter (D_{TSV}) to thickness of silicon (T_{Si}), which in turn is governed by techniques used to etch, deposit the oxide and plating of TSVs. Thus silicon thickness must scale with TSV diameter for fixed aspect ratio. Handling of thin die is more difficult than handling of thin wafers and thus wafer stacking supports smaller TSVs and TSV keep out, and much higher 3D connection density thereby allowing much higher bandwidths. Also, by allowing TSVs to directly land on a wiring level, the need of micro-pillars for 3D communication is eliminated in wafer stacking, thereby overcoming the limitation imposed by limited scaling of micro-pillar pitch, and, thus, favoring further scaling of TSV pitch that supports higher bandwidth.

Two key applications for wafer stacking are envisioned to be (a) massively parallel simple cores; (b) scaling of commodity memory. The increasing cost of lithography, and reliability and yield issues associated with the scaling of technology have made the cost reduction with scaling of technology node very difficult. Wafer scale 3D stacking allows scaling with existing technology nodes. Prior work [3] has presented wafer-scale integration with tungsten TSV for low-power applications.

In this paper we present the stacking of high performance POWER7™ cache cores [4] in 45 nm SOI technology with EDRAM and 13 metal levels. Five micrometer diameter electrically isolated Cu TSVs at 13 μm pitch are used for power delivery and signal communication in the stacked cache cores. Wafers are aligned and joined using low temperature oxide bonding after nine levels of metals. The wafers are thinned to 13 μm using grind polish and etch. TSVs are defined post bonding and thinning using conventional alignment techniques. Up to four additional metal levels are formed post bonding and TSV definition.

The remainder of this paper is organized as follows: Section 2 provides a review of the existing methods for wafer-scale bonding along with a comparison of how the existing methods differ from our method of oxide bonding. The details of our wafer stacking 3D technology and process flow are described in Section 3. Section 4 describes the hardware test results and analysis. Finally, conclusions are presented in Section 5.

2. Previous Work

Wafer-scale bonding can yield important benefits with respect to 3D integration. Among these are micro-scale interconnects (IC) with lower IC delay, very high data bandwidth due to the tighter possible IC pitch, and lower power consumption. In addition, very high throughput is achievable as multiple chips are bonded in parallel fashion and singulated later. As a result, cost savings and high-volume manufacturing can be achieved. With minimal modifications and additions to current manufacturing infrastructure, wafer-scale integration is possible, provided a suitable bonding approach is selected. There are several options regarding the wafer bonding approach, several of which are discussed below.

2.1. Metal-Metal Bonding

This process utilizes metal micro-bumps or pillars/studs on the bonding surface of each wafer and subsequent wafer bonding and it establishes direct electrical connection between the two wafers. There are various methods for achieving the formation of the metallic protrusions that can be used for this. The well-known method of using C4 interconnects, typically used in chip packaging, is not suitable for multiple wafer stacking due to thermal budget limitations and the large size of C4s, which limits bandwidth [5]. Therefore, alternative methods have been considered, as described below.

One example is Au/Sn soldering, which involves micro-bump formation. This utilizes the eutectic compositions of the Au/Sn metallurgy [6]. Such an approach can have some important advantages besides direct electrical connection, such as low soldering temperature, self-aligning during bonding, very good wetting behavior, adequate corrosion resistance, *etc*. However, it may suffer from mechanical stability issues and typically requires under fill processes to ensure mechanical robustness, which increases cost and complexity. In addition, the alignment performance and scaling of interconnections critical dimension (CD) and pitch can be much more challenging, as it is limited by the dimensions of the micro-bumps and throughput may also be a concern.

Another metal-metal wafer bonding method involves the use of solid-liquid inter diffusion (SLID) bonding. In this case, a higher melting point metal (*i.e.*, Cu) is combined with a lower melting point metal (*i.e.*, Sn) [7]. The key in this case is to place the lower melting point metal between the higher melting point metal pads/studs/pillars. Once such structures are created, a thermal compression step is applied in order to facilitate the melting of the low meting point metal and its diffusion into the higher melting point metal with which it forms inter-metallic states. This technique has similar overall advantages and disadvantages to the soldering described above. However, it suffers from limited scalability for multi-wafer stacking due to the limited thermal stability of the bonds *vs.* what is needed for further wafer processing.

Metal-metal thermal compression bonding is a somewhat more promising case for wafer-scale bonding compared to the ones above. In this method, the metal structures (*i.e.*, Cu micro studs/pillars) typically protrude out of the dielectric surface from the wafer surface, usually due to a recess process for the dielectric [8]. It is critical to prevent corrosion and create clean metal surfaces prior to bonding, therefore special clean processes that remove the oxide and other impurities are used to prepare the surface of the wafers. This approach has the advantage of direct electrical connection and the high bond energy of the metal-metal bond interfaces. However, this also typically requires under fill to address mechanical stability and reliability concerns, which increases cost and complexity. The bonding overlay/alignment performance on a wafer-scale needs to be very accurate, especially for small CD interconnect features. In addition, scaling of interconnections CD and pitch can be more challenging *vs.* oxide bonding and may limit the application when aggressive interconnecting schemes are needed, as the alignment performance must be perfect to ensure maximum contact. In addition, the thermal and compression stress can be a challenge for process yield. Finally, throughput is also a concern as the thermal compression bonding is rather lengthy.

2.2. Hybrid Bonding

In this wafer bonding process, bonding of the metallic features on the surface of the wafers (for direct electrical connectivity) are accompanied by the bonding of the surrounding dielectric surfaces, either oxide or polymer, depending on the exact method used [9]. The dielectric can be oxide inter layer dielectric (ILD) or polymer (*i.e.*, BCB). Special surface preparation steps are used, such as chemical mechanical polishing, plasma and/or wet cleans for surface cleaning and surface termination/activation. Typically, low surface topography and atomically smooth roughness is required, although use of recessed metal structures with respect to the dielectric is common. The bonding typically involves an initial room temperature bonding step to preserve alignment, followed by a thermal compression bonding step in the case of polymer/metal hybrid bonding, which ensures bonding of both metal-metal and polymer-polymer surfaces and establishes electrical connectivity.

This approach features the advantage of direct electrical connection and the high bond strength typical in metal-metal bonding interfaces, It does not require under fill, which is an advantage *vs.* metal-metal bonding schemes, in order to address mechanical stability. However, the bonding overlay/alignment performance on a wafer-scale must be very accurate, especially for smaller CD features, and the scaling of interconnections CD and pitch can be more challenging *vs.* oxide bonding and may limit the application when dense interconnecting schemes are needed [10,11]. There is also the question of reliability, as Cu surfaces are not enclosed in impenetrable barriers and are open/bonded to the ILD dielectric (oxide or polymer). Thermal and mechanical stress can be a concern for process yield in the case of polymer/metal hybrid bonding. Overall, throughput can be a concern, especially if thermal compression steps are involved.

2.3. SiO$_2$ Bonding

This wafer bonding process generally involves the formation and preparation of silicon oxide bonding layers on the host and donor wafers and then activation and cleaning of the bonding layers with plasma and aqueous cleaning treatment, followed by loading the wafers, aligning, initiating contact, releasing the wafers to spread the bonding, and finally a post-bonding anneal to promote full-strength covalent bonding between the wafers [12].

Oxide-to-oxide bonding is a highly promising front-up choice due to several significant advantages with regard to manufacturability, flexibility, and reliability. It features stable, insulating bonding layers and interfaces and a high-throughput process when compared to thermal compression bonding. The use of oxide layers provides flexibility with respect to overcoming topography. Bonding alignment performance can be accurate and stable, as initial bonding (which locks the alignment) is at room temperature. In addition to throughput advantages, this technique avoids increased thermal and mechanical stresses that are inherent in thermal compression bonding. Furthermore, as it does not involve direct electrical connections at bonding, it does not present as big a challenge with respect to reliability as metal-metal or hybrid bonding, where misalignment during bonding and/or insufficient under fill process performance can lead to issues, such as copper poisoning, migration/diffusion, mechanical robustness, *etc.* Lastly, oxide bonding is most amenable to further IC scaling of CD and pitch via TSV scaling, which can be much smaller than the typical micro-bump based schemes.

3. Wafer Level 3D Integration Process

High-speed Power7™ L3 processor cache prototype [13], TSV chains, and FET structures originally designed for 3D die stacking were modified for wafer scale bonding. Backside wiring in the die stacking process was mapped to the front side top most level of the thick wafer (Stratum-2) (Figure 1a,b) to provide a landing pad for the 5 μm copper TSVs. The comparison of die stacking and wafer 3D stacking technologies is shown in Figure 1c. In addition to offering smaller TSV and TSV pitch, wafer stacking using oxide bonding compared to micro-bump-based bonding in die stacking. Additionally, die stacking requires additional grind side wiring levels for connection with TSV and pad formation for micro-bump which are no longer required in wafer stacking [14].

Three-hundred-millimeter silicon wafers with 45 nm CMOS devices and 13 wiring levels were stacked similar to Obha *et al.* [15] with three differences. First, the wafers were joined using a low temperature (<400 °C) oxide bonding process instead of polymer adhesive. Second, the thick glass handle wafer used to support the device wafer during thinning was replaced by a thick blank silicon handle wafer. Third, the silicon handle wafer was attached using an oxide bonding step instead of temporary adhesive [12]. After thinning the device wafer (Stratum-1) to 10–12 μm, the handle wafer was removed from the stack using a combination of mechanical grinding, reactive ion etch (RIE), and wet chemical thinning. These steps employ conventional silicon fabrication equipment and processes.

Figure 1. (a) Schematic representation of wafer level 3D stacking; (b,c) Comparison of die-to-die (D2D) and wafer-to-wafer (W2W) 3D stacking technologies.

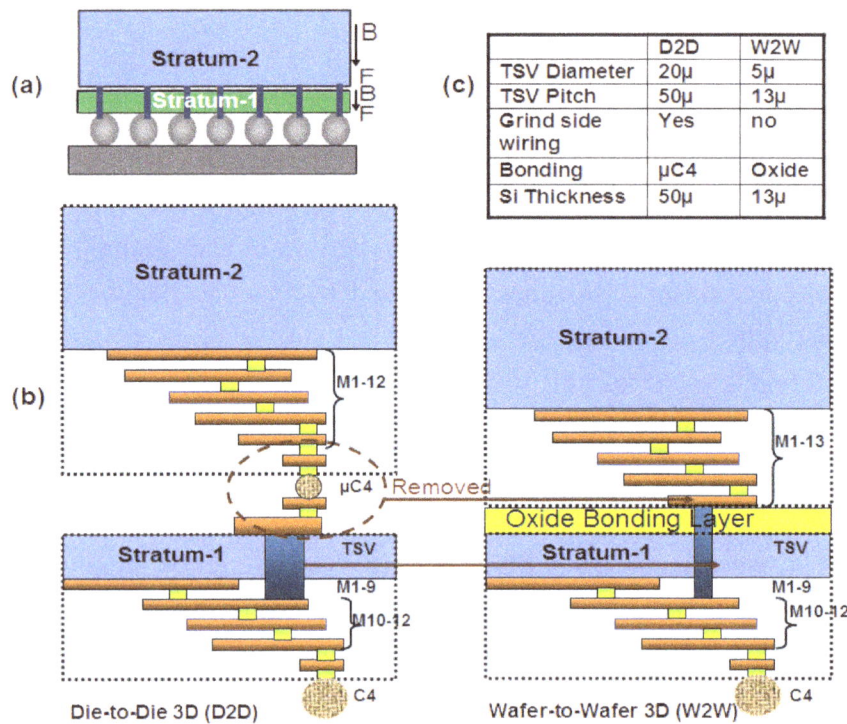

The oxide bonding process used in this work utilizes a dual oxide bonding layer deposited by use of chemical vapor deposition processes. The first film was optimized so as to overcome incoming wafer topography, as bonding surface flatness and absence of short/medium range topography is crucial for defect-free bonding. The film underwent a special annealing step that improves overall cleanliness and density of the surface and enhances the adhesion properties with respect to the second layer. A special polishing step was used to achieve sufficient bonding surface flatness, followed by a special cleaning step. Subsequently, the second bonding layer was deposited, which serves as the primary bonding layer. Special thermal annealing to improve its density and overall cleanliness was used, and a polishing step to ensure suitable roughness that is amenable to oxide bonding [12]. Oxide bonding surfaces characterized using atomic force microscopy (AFM) after the bond film preparation is concluded, are typically shown to be atomically smooth with root mean square (RMS) roughness values between 0.2 nm and 0.4 nm. Thus, a smooth bonding surface is achieved, which is required for the bonding process.

Both host and donor wafers receive the same bonding layer preparation and surface cleaning and then are loaded to the oxide bonding platform where they undergo dual frequency nitrogen plasma activation, followed by de-ionized water megasonics clean. This step is needed to ensure removal of large particles that could result in bonding voids and also terminates the activated hydrophilic surface with silicon to hydroxyl bonds. Once both wafers are activated and cleaned, they are loaded to bonding chucks and they are aligned for bonding. The alignment positions are based on detection of alignment keys patterned on the wafers during the last metal level fabrication. Once the alignment is completed, the wafers are brought in close proximity a few microns from each other. At that point a small piston applies a force on the backside of the top wafer enough to bend it to initiate contact with the bottom wafer. Strong van der Waals bonds between the activated oxide surfaces are formed at the area of contact. After a brief waiting period to lock in the initial bond alignment the top wafer is allowed to relax on the surface of the bottom wafer and a bonding wave from the initially bonded area at the center propagates to the edge of the wafers. Then, the bonded wafer pairs are transferred without bonding overlay alignment deterioration to an anneal furnace, where the anneal can promote strong full-covalent bonding.

For the purposes of this work, a blank silicon wafer was bonded to the device wafer for use as handler. The silicon handler wafer was attached using the low-temperature oxide bonding method. The device wafer was then thinned from the backside to a thickness of 10–13 μm by using a combination of mechanical grinding, reactive ion etch (RIE), and wet chemical thinning, all these processes employing conventional silicon fabrication equipment and processes. An oxide layer was then prepared on the backside of the device wafer and then the stack was bonded to another paired full-thickness device wafer, with the blanket handler wafer now at the top of the stack. The handler wafer was then removed from the stack, also by a combination of mechanical grinding, reactive ion etch (RIE), and wet chemical thinning.

After stacking and removal of the handler, 5 μm diameter copper TSV interconnects were formed by use of deep RIE etching from the front side of stratum-1 (Figures 2 and 3) using a process designed to handle the complex interlayer dielectric stacks, bond interface, and thinned silicon. The TSVs were lined with conformal oxide insulator liner by sub-atmospheric chemical vapor deposition (SACVD). Another RIE step was used to open the TSV insulator at the bottom of the TSVs. This was followed by TSV metallization, which involves metal liner and Cu seed deposition with physical vapor deposition (PVD) methods and finally bottom-up copper plating to fill the TSVs. In this fashion, TSVs with critical dimensions (CD) at 5 μm at a 13 μm pitch were fabricated.

Three additional copper wiring levels were then fabricated on stratum-1. Low resistivity copper TSVs integrated with copper wiring is essential to limit IR voltage drop for high-performance and low-voltage applications [2,16]. All processing and tools are compatible with advanced metal gate down to at least the 14 nm technology node and the TSV diameter is scalable to 1 μm. The face-to-back process can be repeated to stack multiple wafers.

Figure 2. SEM for stratum-1 and stratum-2 after bonding and removal of handle wafer.

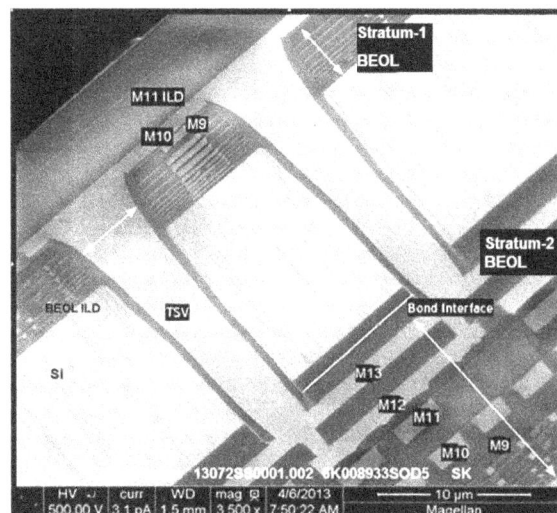

Figure 3. Cross-section SEM showing integrated TSV and 25 BEOL structures (45 nm).

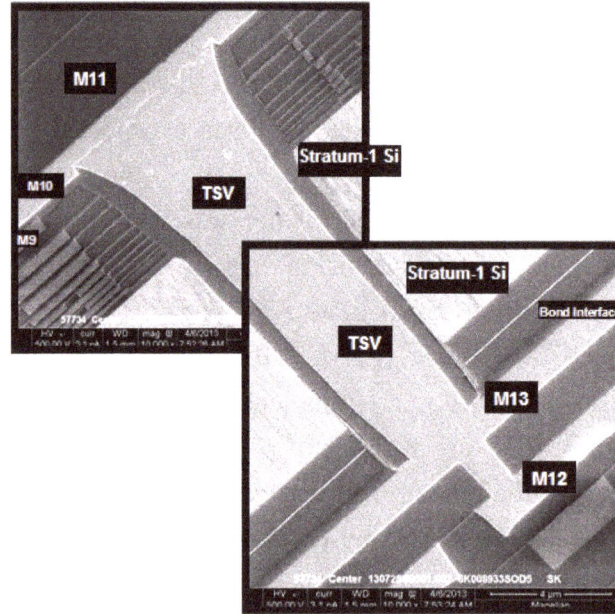

4. Hardware Results

The resistance of TSV chain structures, each link containing two TSVs in parallel, indicates 65 mΩ/link including TSV and local wire resistance (Figure 4). This can support high performance applications requiring over 1 Amp/mm^2 current density while controlling TSV area penalty under 1%. The measured resistance (Figure 4b) has standard deviation of 10 mΩ/link indicating a controlled TSV process. TSV capacitance is measured at ~40 fF (Figure 5) which is <1/4 of the TSV capacitance in bump bond technology [2,14]. The TSV RC characteristic is similar to typical 100–200 µm length wiring load enabling 3D macro-to-macro signaling without additional buffering. Capacitance (Figure 5) and leakage to substrate of TSV arrays (Figure 6a) modulates with the number of TSVs. The leakage is near the tester limit and extrapolates to 1.18 pA/TSV at 2 V. For a 400 mm^2 chip having up to 50 K TSVs, the total leakage due to TSVs is estimated to be 59 nA (Figure 6b) which is negligible for most applications. FET Ion/Ioff shows no significant change post stacking and TSV processing (Figure 7).

Figure 4. (**a**) Measured TSV chain resistances for lengths of 12, 22, 46, 72, and 82; (**b**) Distribution of measured resistance/link showing σ = 10 mΩ/link.

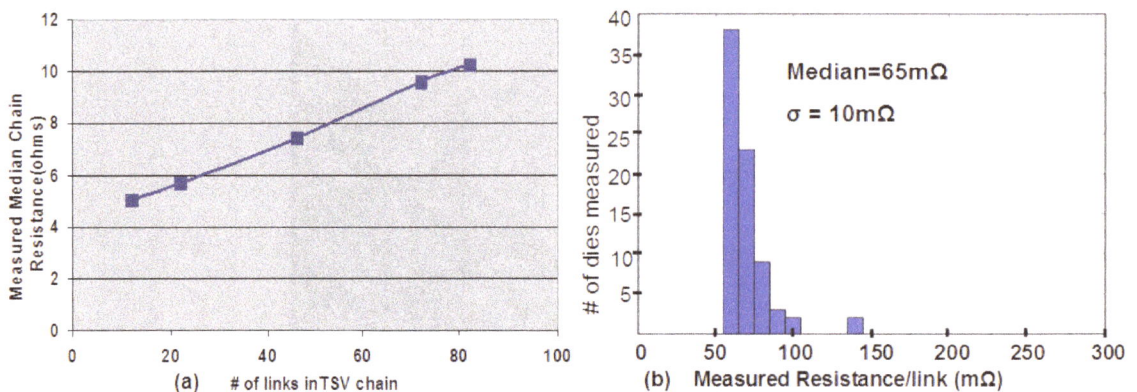

Figure 5. Capacitance for TSV banks of 80, 160, 240, 320 TSVs.

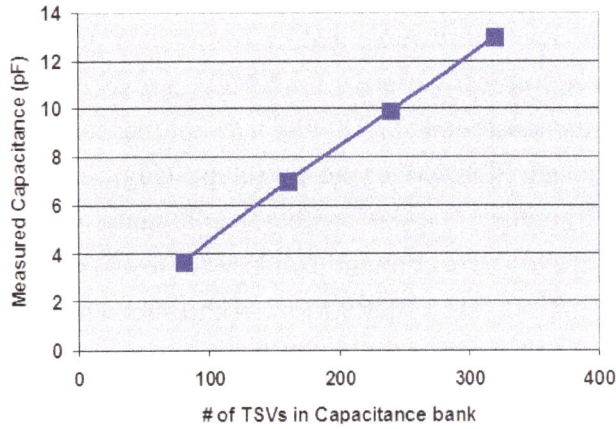

Figure 6. (**a**) TSV Leakage to substrate at 2 V measured in TSV arrays having 80, 160, 240, 320 TSVs; (**b**) Distribution of leakage current through each TSV across multiple wafers and dies.

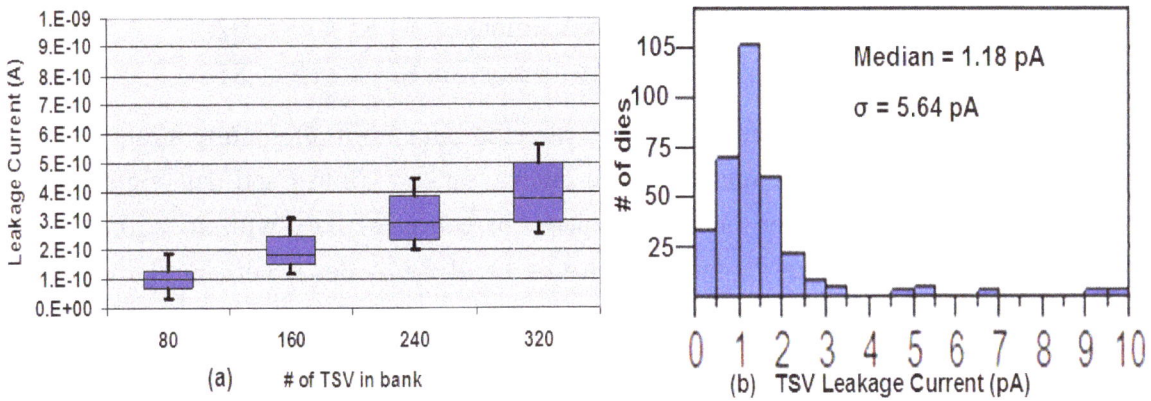

Figure 7. PMOS (**a**) Idlin and (**b**) Ioff distribution post bonding, thinning and TSV fabrication.

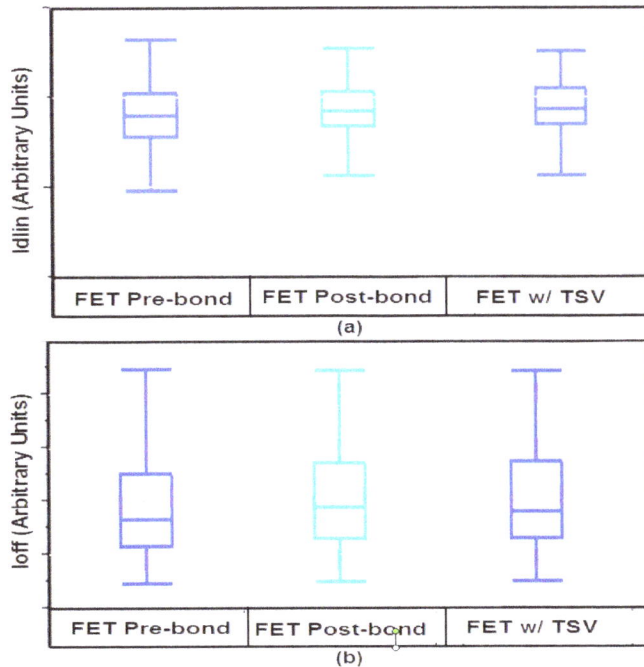

As shown in Figure 7, no significant change in PMOS Idlin and Ioff was observed compared with Ion/Ioff of 2D devices. Pre-bond measurement is done on thick wafers before bonding, post-bond on stratum-1 after bonding before TSV fabrication, and FET with TSV is measured after TSV fabrication on stratum-1. The TSV induced Ion/Ioff variation for NMOS transistors is expected to be smaller than PMOS devices as previously demonstrated [17,18] and was not measured for this work.

The high-performance stacked EDRAM cache prototype (Figure 8a) was built using more than 11,000 TSVs. Strata-1 and 2 emulate stacked processor and cache respectively as described in [5]. Memory functionality and strata-to-strata communication were tested using the built-in-self-test engine (BIST) on each strata. The BIST accessed EDRAM on both strata-1 and 2. The shmoo plot of supply voltage and frequency (Figure 8b) shows 16 Mb EDRAM functionality (fixable) and strata-to-strata communication up to 2.1 GHz at 1.3 V.

The memory patterns were written in the stacked EDRAM in four different configurations:

(a) 2D thick wafer mode where the memory on the thick S1 wafer was activated;

(b) 2D thin wafer mode where the memory on thin wafer S1 was activated and the test patterns were loaded using the TSVs;

(c) 3D mode where the BIST on S1 controls the memory on S1 as well as S2;

(d) 3D mode where the BIST on S2 controls the memory on S2 as well as S1.

Modes (c) and (d) demonstrate the ability to write/read data from alternating strata memory in a single cycle thereby confirming the quality of the clocks, power, control and data signals across the chip boundary to be able to transfer the data at speed for entire memory with no errors. Further the shmoo plots in Figure 8b show that the failure signature in all the modes is very similar indicating that the EDRAM behavior was unaffected by the 3D processing and TSVs. The pattern shmoo was the march9 pattern that forces cycle-to-cycle simultaneous switching patterns across strata boundary. An equivalent column march pattern was also run and the results were similar. The maximum frequency at wafer test was limited by voltage drop inherent with cantilever probing compared to socket based module test. Retention signature of EDRAM indicates a retention time of over 200 μs.

Figure 8. (a) 3D EDRAM on each wafer comprising 3–16 Mb IP blocks; (b) Shmoo plot showing performance characteristics of wafer-to-wafer stack EDRAM tested with a cantilever wafer probe.

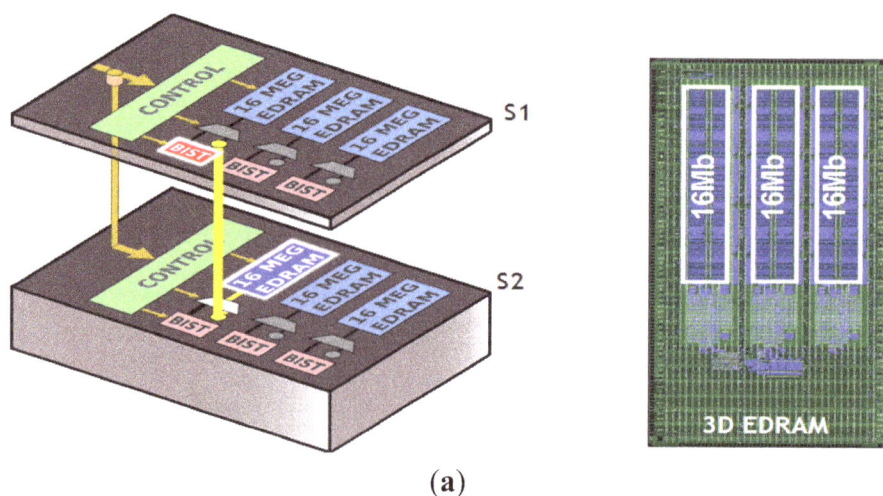

(a)

Figure 8. *Cont.*

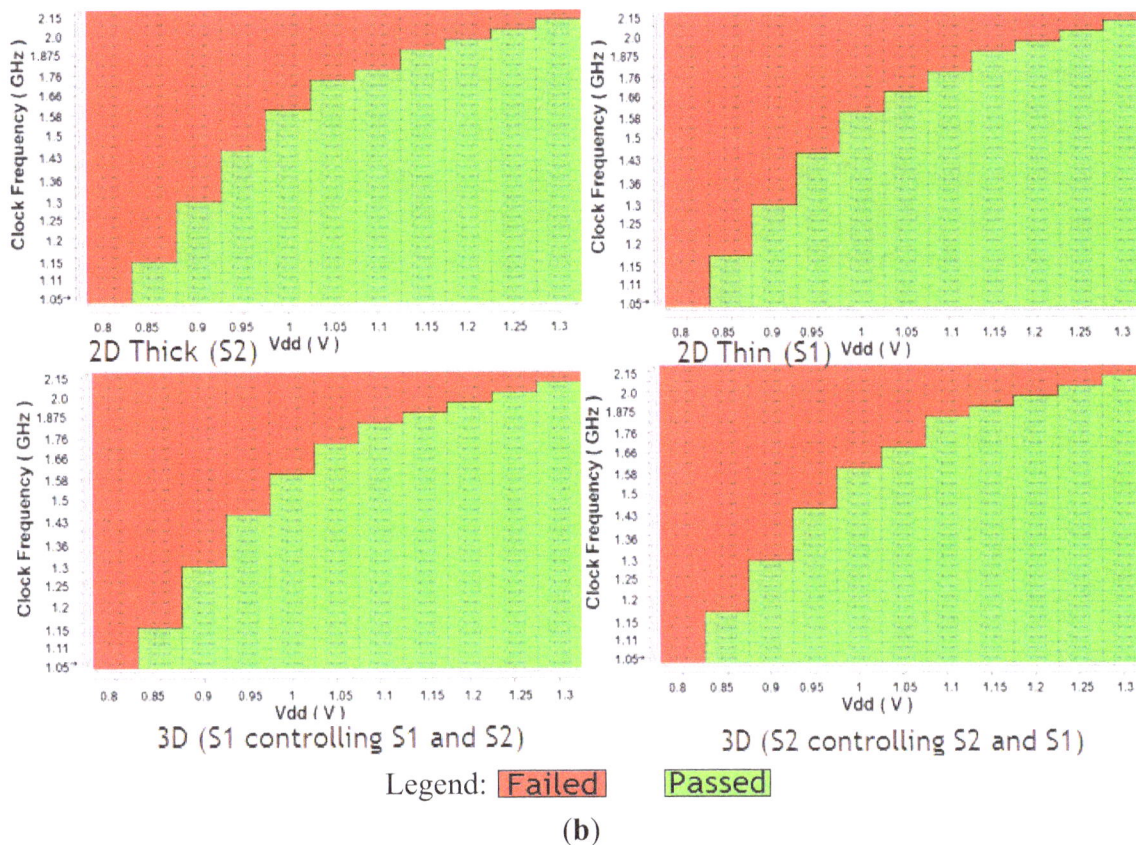

2D Thick (S2)

2D Thin (S1)

3D (S1 controlling S1 and S2)

3D (S2 controlling S2 and S1)

Legend: Failed   Passed

(b)

5. Conclusions

3D technology for wafer scale stacking using oxide bonding has been demonstrated in 45 nm SOI-CMOS EDRAM technology. TSV capacitance and resistance (including wire) was measured to be 40 fF and 65 mΩ, suitable for high bandwidth chip-chip communication. TSV leakage of 1 pA/TSV is negligible. FET Ion/Ioff shows no significant change post stacking and TSV processing and functionality of 3D stacked EDRAM cache cores has been confirmed by successful writing of memory patterns at up to 2.1 GHz at 1.3 V. The compatibility with existing high performance logic technology makes this wafer scale stacking process ideal for volume production of 3D scaled logic and memory applications. Future focus items for this technology include further reduction of the TSV size and pitch and extending the number of stacked wafers to four and higher.

Acknowledgments

The authors would like to thank Matthew R. Wordeman, Michael R. Scheuermann, Joel A. Silberman and Christy S. Tyberg at Watson Research Center.

Conflicts of Interest

The authors declare no conflict of interest.

References

1. Iyer, S.S.; Kirihata, T.; Wordeman, M.R.; Barth, J.; Hannon, R.H.; Malik, R. Process-Design Considerations for Three Dimensional Memory Integration. In Proceedings of the Symposium on VLSI Technology, Honolulu, HI, USA, 16–18 June 2009; pp. 60–63.

2. Farooq, M.G.; Graves-Abe, T.L.; Landers, W.F.; Kothandaraman, C.; Himmel, B.A.; Andry, P.S.; Tsang, C.K.; Sprogis, E.; Volant, R.P.; Petrarca, K.S.; *et al.* 3D Copper TSV Integration, Testing and Reliability. In Proceedings of the IEEE International Electron Devices Meeting (IEDM), Washington, DC, USA, 5–7 December 2011.

3. *Wafer Level 3-D ICs Process Technology*; Tan, C.S., Gutmann, R.J., Reif, R., Eds.; Springer US: New York, NY, USA, 2008; p. 286.

4. Barth, J.; Reohr, W.R.; Parries, P.; Fredeman, G.; Golz, J.; Schuster, S.E.; Matick, Richard E.; Hunter, H.; Tanner, C.C.; Harig, J.; *et al.* A 500 MHz random cycle, 1.5 ns latency, SOI embedded DRAM macro featuring a three-transistor micro sense amplifier. *IEEE J. Solid-State Circuits* **2008**, *43*, 86–95.

5. Totta, P.A.; Khadpe, S.; Koopman, N.G.; Sheaffer, M.J. Chip-to-Package Interconnections. In *Microelectronics Packaging Handbook, Part II*, 2nd ed.; Tummala, R.R., Rymaszewski, E.J., Klopfenstein, A.G., Eds.; Springer: New York, NY, USA, 1997; pp. 129–283.

6. Oppermann, H.; Hutter, M. Au/Sn Solder. In *Handbook of Wafer Bonding*; Ramm, P., Lu, J.J.-Q., Taklo, M.M.V., Eds; Wiley-VCH Verlag GmbH & Co.: Weinheim, Germany, 2012; p. 119.

7. Hoivik, N.; Aasmundtveit, K. Wafer Level Solid-Liquid Interdiffusion Bonding. In *Handbook of Wafer Bonding*; Wiley-VCH Verlag & Co.: Weinheim, Germany, 2012; p. 181.

8. Chen, K.N.; Tan, C.S. Thermocompression Cu-Cu Bonding of Blanket and Patterned Wafers. In *Handbook of Wafer Bonding*; Wiley-VCH Verlag & Co.: Weinheim, Germany, 2012; p. 161.

9. Di Cioccio, L. Cu/SiO$_2$ Hybrid Bonding. In *Handbook of Wafer Bonding*; Wiley-VCH Verlag & Co.: Weinheim, Germany, 2011; p. 237.

10. Ko, C.-T.; Hsiao, Z.-C.; Fu, H.-C.; Chen, K.-N.; Lo, W.-C.; Chen, Y.-H. Wafer-to-Wafer Hybrid Bonding Technology for 3D IC. In Proceedings of the Electronic System-Integration Technology Conference, Berlin, Germany, 13–16 September 2010; pp. 1–5.

11. Ko, C.-T.; Chen, K.-N.; Lo, W.-C.; Cheng, C.-A.; Huang, W.-C.; Hsiao, Z.-C.; Fu, H.-C.; Chen, Y.-H. Wafer-Level 3D Integration Using Hybrid Bonding. In Proceedings of the IEEE International 3D Systems Integration Conference, Munich, Germany, 16–18 November 2010; pp. 1–4.

12. Skordas, S.; Tulipe, D.C.L.; Winstel, K.; Vo, T.A.; Priyadarshini, D.; Upham, A.; Song, D.; Hubbard, A.; Johnson, R.; Cauffman, K.; *et al.* Wafer-Scale Oxide Fusion Bonding and Wafer Thinning Development for 3D Systems Integration. In Proceedings of the 3rd IEEE International Workshop on Low Temperature Bonding for 3D integration (LTB-3D), Tokyo, Japan 22–23 May 2012; pp. 203–208.

13. Wordeman, M.; Silberman, J.; Maier, G.; Scheuermann, M. A 3D System Prototype of an EDRAM Cache Stacked Over Processor-Like Logic Using Through-Silicon Vias. In Proceedings of the IEEE International Solid State Circuits Conference (ISSCC), San Francisco, CA, USA, 19–23 February 2012; pp. 186–187

14. Golz, J.; Safran, J.; He, B.; Leu, D.; Yin, M.; Weaver, T.; Vehabovic, A.; Sun, Y.; Cestero, A.; Himmel, B.; *et al.* 3D stackable 32 nm High-K/Metal Gate SOI embedded DRAM prototype. In Proceedings of the Symposium on VLSI Circuits, Honolulu, HI, USA, 15–17 June 2011; pp. 228–229.

15. Ohba, T. Bumpless WOW Stacking for Large-Scale 3D Integration. In Proceedings of the 10th IEEE International Conference on Solid-State and Integration Circuit Technology (ICSICT), Shanghai, China, 1–4 November 2010; pp. 70–73.

16. Sarvari, R.; Naeemi, A.; Zarkesh-Ha, P.; Meindl, J.D. Design and Optimization for Nanoscale Power Distribution Networks in Gigascale Systems. In Proceedings of the IEEE International Interconnect Technology Conference, Burlingame, CA, USA, 4–6 June 2007; pp. 190–192.

17. Mercha, A.; Van der Plas, G.; Moroz, V.; de Wolf, I.; Asimakopoulos, P.; Minas, N.; Domae, S.; Perry, D.; Choi, M.; Redolfi, A.; *et al.* Comprehensive Analysis of the Impact of Single and Arrays of Through Silicon Vias Induced Stress on High-K/Metal Gate CMOS Performance. In Proceedings of the IEEE International Electron Devices Meeting (IEDM), San Francisco, CA, USA, 6–8 December 2010; pp. 2.2.1–2.2.4.

18. Yu, L.; Chang, W.-Y.; Zuo, K.; Wang, J.; Yu, D.; Boning, D. Methodology for Analysis of TSV Stress Induced Transistor Variation and Circuit Performance. In Proceedings of the 13th International Symposium on Quality Electronic Design (ISQED), Santa Clara, CA, USA, 19–21 March 2012; pp. 216–222.

8

The Impact of Process Scaling on Scratchpad Memory Energy Savings [†]

Bennion Redd [1,*], Spencer Kellis [2], Nathaniel Gaskin [3] and Richard Brown [1]

[1] Department of Electrical & Computer Engineering, University of Utah, 1692 Warnock Engineering Bldg., 72 S. Central Campus Dr., Salt Lake City, UT 84112, USA; E-Mail: brown@utah.edu

[2] Division of Biology and Biological Engineering, California Institute of Technology, Mail Code 216-76, 1200 E, California Blvd., Pasadena, CA 91125, USA; E-Mail: skellis@vis.caltech.edu

[3] 701 E Charleston Rd, Palo Alto, CA 94303, USA; E-Mail: gaskinnc@gmail.com

[†] The manuscript is an extended version of a conference paper titled "Scratchpad Memories in the Context of Process Scaling", which was presented at the 2011 IEEE 54th International Midwest Symposium on Circuits and Systems (MWSCAS).

[*] Author to whom correspondence should be addressed; E-Mail: bennion.redd@utah.edu

Abstract: Scratchpad memories have been shown to reduce power consumption, but the different characteristics of nanometer scale processes, such as increased leakage power, motivate an examination of how the benefits of these memories change with process scaling. Process and application characteristics affect the amount of energy saved by a scratchpad memory. Increases in leakage as a percentage of total power particularly impact applications that rarely access memory. This study examines how the benefits of scratchpad memories have changed in newer processes, based on the measured performance of the WIMS (Wireless Integrated MicroSystems) microcontroller implemented in 180- and 65-nm processes and upon simulations of this microcontroller implemented in a 32-nm process. The results demonstrate that scratchpad memories will continue to improve the power dissipation of many applications, given the leakage anticipated in the foreseeable future.

Keywords: scratchpad memory; loop cache; process scaling; low power; microprocessor; computer architecture; embedded

1. Introduction

Modern microprocessors may use techniques, such as small low-power scratchpad memories or very small instruction caches, to reduce power consumption. Both of these types of memory reduce the average power of on-chip memory accesses by making the most-frequent accesses to these small memories. The scratchpad memory uses even less energy than a cache, because it does not have tag storage or comparison circuits. These techniques have been demonstrated to reduce power consumption in older technologies, but their benefits need to be reevaluated for new nanometer-scale processes. Higher leakage power in these processes could outweigh the energy savings gained by adding a scratchpad memory, so the assumptions made in previous analyses need to be reexamined.

This paper describes a model that quantifies the benefits of scratchpad memories using memory access rates, access energies and leakage power data. The model is used, along with data from CACTI (a cache and memory access time model) simulations, to show how both process technology and application characteristics impact the energy savings resulting from the inclusion of a scratchpad memory. The concepts demonstrated by this model put the importance of leakage power into perspective. In this expanded version of our earlier analysis [1], leakage trends are evaluated using Intel's process technology reports to see whether scratchpad memories will be viable power-saving architectural features in the foreseeable future.

Measured data from the WIMS (Wireless Integrated MicroSystems) microcontroller [2] can be used to verify this conclusion, because this microcontroller has a scratchpad memory and has been implemented in both 180- and 65-nm processes. The energy profile of a 32-nm version of the microcontroller can be projected using reported Intel transistor data. Energy savings due to the incorporation of a scratchpad memory will be projected for several benchmark programs using the previously described model and measured data from the WIMS microcontroller.

2. Background

Scratchpad memories are small, low-power memories that can reduce power consumption by providing low-energy access to frequently used data or program fragments, such as loops. These memories are most effective for applications that are dominated by many small loops or that have small data blocks that are frequently accessed [3].

One precursor to the WIMS approach to scratchpad memories was the employment of a very small direct-mapped instruction cache, called a filter cache [4]. It reduced access energy, but was still subject to thrashing and conflicts. This filter cache was refined into a loop cache (LC) by using information from a profiler to allow the compiler to place instructions into the loop cache optimally. An instruction was added to inform the processor of which blocks should be placed into the loop cache [5]. Other approaches recognized the loop to be placed into the loop cache from a new "short backward branch" instruction, avoiding the need for profiling or tag comparison circuitry in the LC [6,7].

To overcome the burden of added hardware complexity in the loop cache, a scratchpad memory (a memory without a hardware controller) was developed. In this approach, the compiler chooses frequently used data, or code fragments [3], or virtual memory pages [8] to statically reside in the scratchpad memory. These objects are accessed by an absolute memory address and are software managed, which reduces hardware complexity by eliminating both the tag comparison circuitry and the LC controller. When the most frequently accessed objects are placed in the scratchpad memory, less energy is used by the system than would have been used with either the filter cache or LC approaches [9]. Profiling is used in this case.

The WIMS microcontroller compares two scratchpad memory approaches: a static approach, in which code or data blocks in the scratchpad memory remain constant for a compiled program; and a dynamic approach, in which the contents of the scratchpad memory change over the course of the execution of a program [10]. These approaches are explained in more detail in Section 6.3.

3. Motivation

Early research on scratchpad memories was conducted on pre-nanometer processes [4–7,9]. As semiconductor processes are scaled, feature size reduction and material changes affect power consumption and memory performance, which could affect the benefits of scratchpad memory. In particular, increased leakage power could outweigh the dynamic power savings of scratchpad memories. Researchers have anticipated [11] that as processes are scaled, leakage will become a greater problem.

Recent research on scratchpad memory has addressed the increasing importance of leakage power by finding ways to minimize leakage. Guangyu *et al.* [12] split the scratchpad memory into different banks and considered turning off one or more banks for some loop nests to minimize leakage energy. Each loop nest was analyzed using integer linear programming to determine the optimal number of scratchpad memory banks to use. Kandemir *et al.* [13] mapped data with similar temporal locality to the same scratchpad memory bank to maximize bank idleness and to increase the chances of that bank entering a drowsy state, thereby decreasing power. Huangfu *et al.* added compiler instructions, so that their compiler can activate and deactivate groups of words in the scratchpad memory to significantly reduce leakage without a significant performance penalty [14]. Takase *et al.* [15] used data on leakage trends from CACTI 5.0 [16] to consider how leakage trends impact scratchpad memories across two process transitions.

The study described in [1] added measured data from the WIMS microcontroller implemented in two process generations, another simulated process transition (for a total of three) and a model to assist in characterizing measured data from a fabricated processor. The present paper expands upon that study by examining more recent data on leakage trends from industry transistor reports from SoC processes, energy savings projections for the WIMS microcontroller based on the most recent transistor reports and information on the more effective dynamic scratchpad memory allocation algorithm, which was implemented for the WIMS microcontroller.

4. Conceptual Framework

4.1. Models

To clarify the relationship between energy savings, access rates (the percentage of total accesses that are directed to the scratchpad memory) and access energies, a simple mathematical model was developed.

Dynamic energy savings (ES) are calculated from scratchpad access rates (AR), scratchpad fetch energy (SFE) and the main memory fetch energy (MFE):

$$1 - ES = \frac{(AR \cdot SFE + (1 - AR) \cdot MFE)}{MFE} \tag{1}$$

This model applies to both systems with on-chip and off-chip main memories. The denominator of the fraction on the right-hand side of (1) represents energy without a scratchpad memory, and the numerator represents energy usage with a scratchpad memory. Simulations presented in [10] for the 180-nm process assumed the model described by (1).

To account for leakage energy, (1) was updated to include the two new variables SLE (scratchpad leakage energy) and MLE (main memory leakage energy):

$$1 - ES = \frac{(AR \cdot SFE + (1 - AR) \cdot MFE) + MLE + SLE}{MFE + MLE} \tag{2}$$

SLE and MLE are defined to be the average leakage energy per memory access, due to the scratchpad memory and main memory, respectively. This equation assumes that the scratchpad and main memory latencies are equal, which is pessimistic for many systems, but accurate for the WIMS microcontroller, which has an on-chip main memory. This model predicts that energy savings will be positive any time the numerator on the right-hand side of the equation is smaller than the denominator. This condition, along with algebraic simplification, results in Equation (3), which indicates whether the scratchpad memory saves power:

$$SLE < AR \cdot (MFE - SFE) \tag{3}$$

These equations will be used with data generated by the CACTI 5.3 Pure RAM Interface [16] to demonstrate how to evaluate the benefits of the scratchpad memory and to show which scaling trends can be used to predict the future benefit of a scratchpad memory. The CACTI-generated SRAMs described by Figures 1–3 have one bank, one read/write port and no other ports and read out 16 bits. These memories are optimized for low standby current. In Section 6, these equations will use measured data from the WIMS microcontroller to project energy savings for several benchmarks.

Figure 1. Leakage power divided by total memory power for different access rates. Data is generated using CACTI (a cache and memory access time model) 5.3 for 65-nm memories optimized for low standby current [16].

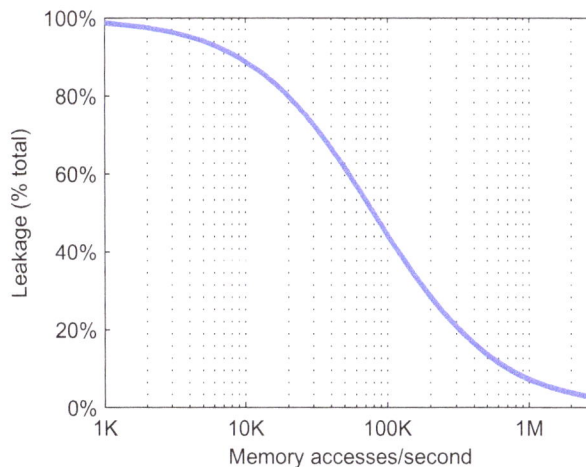

Figure 2. Access energies of low standby power memories of different sizes in different process technologies. Data is generated using CACTI and is based on ITRS (the International Technology Roadmap for Semiconductors) 2005 projections.

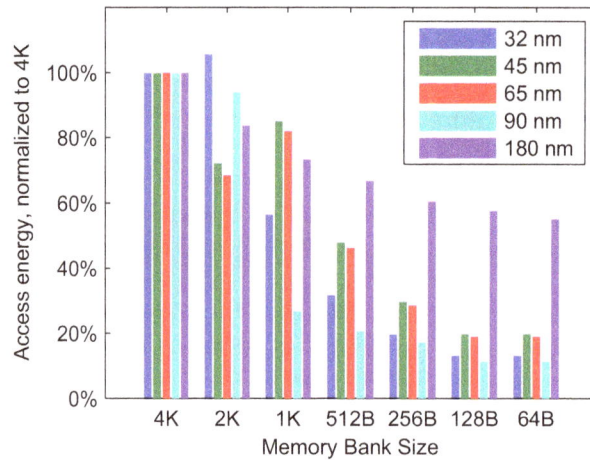

Figure 3. Leakage energy per access cycle for different processes with the maximum number of memory accesses for the process (blue) and a fixed number of memory accesses per second lower than the maximum frequency of any of the examined processes (green). Arrows indicate the general scaling trend of each type of application.

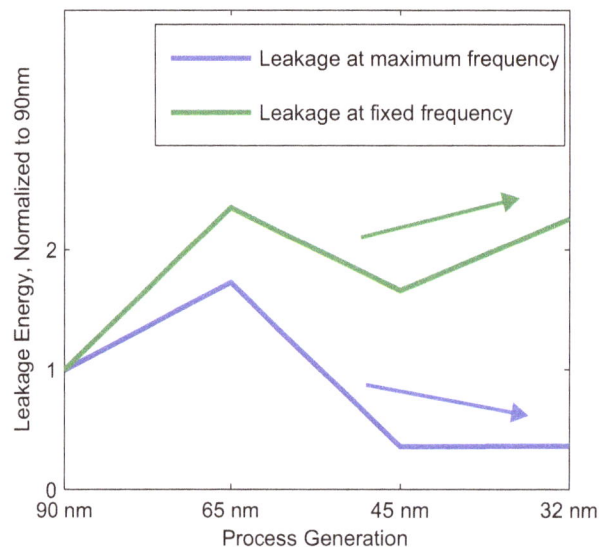

4.2. Relative Importance of Active and Leakage Energy

A small memory, such as a scratchpad memory, will usually have lower access energy than a large memory, like the main memory. On the other hand, the addition of a scratchpad memory will increase memory leakage power. If the scratchpad memory is used to reduce power, the reduction in access energy must outweigh the increase in leakage power. The scratchpad memory is usually very small compared to the main memory, so the increase in leakage power is generally very small, as well, but may be larger in future processes [11,17]. The relative importance of the access energy compared to the leakage energy depends on how frequently the memory is accessed in a given application. This concept is illustrated by Figure 1, which uses data from CACTI 5.3 to show leakage power as a percentage of total power for a given number of memory accesses per second. This figure demonstrates that for applications that rarely access the memory, leakage energy can dominate the memory's energy consumption, and the additional leakage energy due to the scratchpad memory can be more significant than its access energy savings.

4.3. Access Energy and Memory Size

The benefit of the scratchpad memory, in terms of active energy reduction, is determined by the difference between SFE and MFE. Therefore, a key assumption of scratchpad memories is that small memories have lower access energies than large memories. To verify this assumption, access energy data was collected from CACTI cache simulations: an older version of CACTI [18] for 180 nm and CACTI 5.3 [16] for 90, 65, 45 and 32 nm. CACTI 5.3 uses technology information projected from the 2005 ITRS [16,17]. CACTI projections are based on a real physical layout, which includes detailed device and wire models. Even if device characteristics change over time, if the layout structures are similar, the trend of smaller memories consuming less power should be consistent. The data from CACTI 5.3 are more reliable than from the older version of CACTI, because its models are more closely derived from real physical structures. The normalized access energies of 64 B to 4 KB low standby power memories for these processes are shown in Figure 2. The simulated measurements show that access energy generally decreases with memory size in all of the simulated processes. There are a few exceptions to this rule, because CACTI simulates real layouts for memories that are primarily optimized for area and latency, rather than access energy. However, the general rule that access energy decreases with memory size is expected to hold, because larger memories will usually have longer wires and higher switching capacitance than small memories, so access energy is expected to decrease with memory size in all processes.

4.4. Leakage Energy and Process Scaling

An increase in total leakage current is the primary drawback of a scratchpad memory, so it is important to evaluate how scaling affects leakage energy to predict how beneficial scratchpad memories will be in future processes. The evaluation of leakage energy scaling trends depends on how often a memory will be accessed for a particular application. The higher frequencies enabled by newer processes help to decrease leakage energy per clock cycle, because the clock cycles are shorter, but if an application only requires a fixed number of memory accesses per second (lower than the maximum number of accesses), then the maximum speed of the memory is irrelevant. An application that accesses memory fewer times per second than the maximum possible frequency might be equivalently viewed as having a fixed memory access frequency lower than the maximum. The leakage trend for such an application running at a frequency fixed across process generations (and below the maximum of any of those processes) is contrasted to a memory running at maximum frequency of each process in Figure 3. The trend for an application with a frequency that is fixed across process generations does not improve as much with process scaling as an application that takes advantage of the maximum process frequency.

The purpose of Figure 3 is only to demonstrate the concept that leakage trends for fixed frequency applications are worse than trends for applications running at maximum frequencies if the newer processes target higher frequencies. Figure 3 is not intended to project specific leakage trends, because the 65-nm and later nodes had not yet been released when the source of its data, the 2005 ITRS, was published. A better description of recent leakage trends will be made in Section 5 using more recently published foundry process reports on transistor characteristics. The CACTI simulator used to generate Figure 3 has not yet been updated using the newest transistor data, so leakage trends must be projected using transistor data instead of simulations of memories.

5. Scaling Trends

Processes developed specifically for SoC applications have lower-power transistors and a larger variety of process options. Because embedded systems that are likely to benefit from a scratchpad memory are also likely to be built into these processes, these are the processes whose transistor characteristics should be examined.

For simplicity, only Intel's processes will be presented. Similar figures could be created for other foundries with low-power SoC processes, but the results would be similar. Figure 4 shows the reported achieved ID_{sat} and I_{off} characteristics for the minimum channel length of the lowest available power type of both NMOS and PMOS transistors in Intel's 65-, 45-, 32- and 22-nm low-power processes [19–22]. These are low-power SoC variants of high-performance processes. The 65-nm node was Intel's first process node to introduce an SoC process. The ID_{sat} parameter reflects the drive strength of the transistor: larger I_{on} corresponds to a faster, higher-performance transistor in saturation; lower I_{off} corresponds to a less leaky, more power-efficient transistor. The standard input voltages have decreased over time to decrease power consumption.

Figure 4. NMOS (**left**) and PMOS (**right**) reported drive currents and off-state leakages for minimum-length transistors of the lowest available power type available in each of Intel's 65-, 45- and 32-nm SoC processes [19–22]. **ID$_{sat}$** is the saturation drain current, or the current a transistor can provide when it is fully on. **I$_{off}$** is the leakage current that is present when a transistor is turned off.

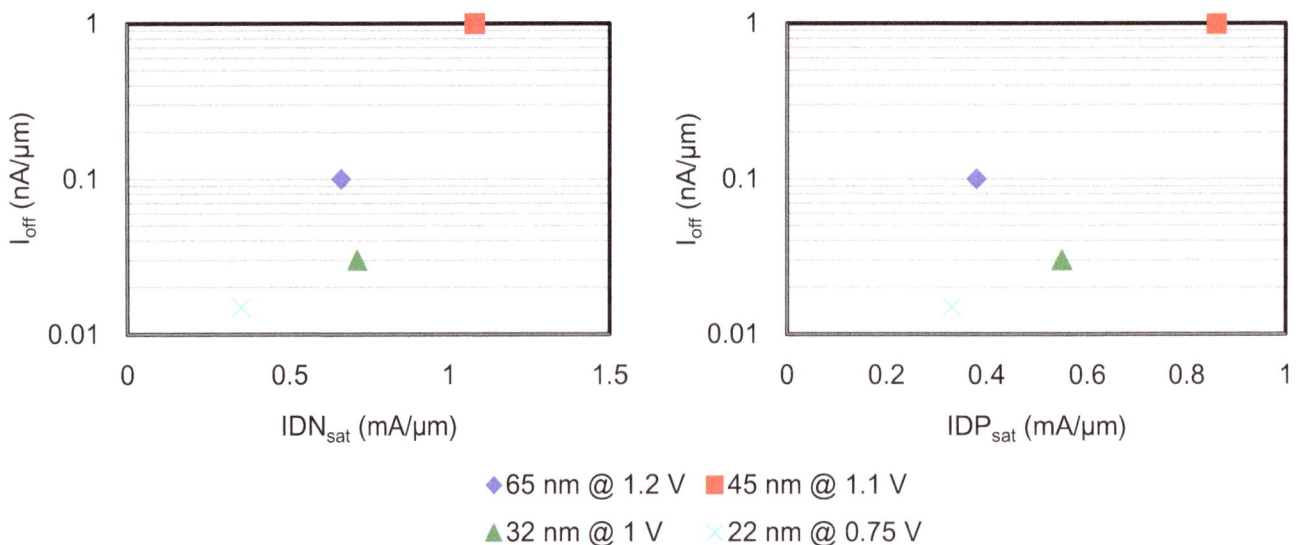

The overall trends illustrated in Figure 4 show that leakage has decreased over time. This decrease can be attributed to many process innovations, which will be examined in detail.

5.1. Process Innovations

Constant field scaling [23] has not been followed faithfully, but over recent process generations, voltages have been reduced. The last several process generations have employed a variety of innovative techniques to improve performance, while reducing leakage. An examination of the variety of process innovations that have driven improvement over the last several process generations indicates that future processes may also improve performance and reduce leakage through further innovations.

Contrary to constant field scaling rules, Intel's 65-nm SoC process increased gate oxide thickness to decrease gate leakage. The good doping profile and source/drain spacers were optimized to prevent an increase in source-drain leakage, but this also increased the threshold voltage and decreased performance. The performance decrease was mitigated by uni-axial strained silicon [19].

The 45-nm SoC process introduced a dramatic structural change: a Hafnium-based high-K metal gate (HKMG), which increased performance and decreased gate leakage by allowing for a physically thicker gate with the equivalent capacitance of a thin gate. Figure 5 shows how dramatically an HKMG can reduce gate leakage. However, gate leakage is still much less than source-drain leakage, and Figure 4 indicates that although the HKMG was introduced at the 45-nm node; total I_{off} dropped much more significantly at the 32-nm node. The 32-nm node also utilized the HKMG and was the first Intel process node to use immersion lithography for the critical layer, allowing for resolution enhancement of that layer [20,21].

Figure 5. The addition of a high-K metal gate decreases leakage across all voltages in an IBM 32-nm process. Taken with permission from [24].

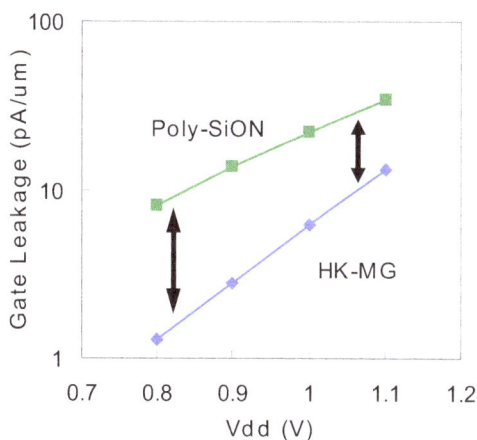

Intel's 22-nm process introduced a significant advancement in process technology: a tri-gate transistor, also known as a FinFET. These transistors have a much steeper subthreshold slope than the planar 32-nm transistors [22]. This steeper subthreshold slope could be used to either significantly reduce leakage energy or to target a lower threshold voltage to increase performance. As with strained silicon and HKMG innovations, as the tri-gate transistor structure is refined, its impact will increase over time.

Research indicates that these transistor improvements are carrying over to SRAM improvements. For example, in the 22-nm technology, Intel created a low standby power SRAM with only 10-pA/cell standby leakages [22].

Other foundries have proposed other innovations for transistor structures that are targeted at reducing leakage energy, such as partially or fully-depleted SOI (silicon on insulator) or SuVolta's deeply depleted channel (DDC) transistor [25]. In new process technologies, the improvements to energy savings and performance are now as much a result of structural redesign and innovation as they are of device dimension and voltage scaling.

5.2. SRAM Cell Design

Even if the process trends do not continue to reduce leakage power, as predicted in the previous section, SRAM cell design can be modified to reduce leakage. The traditional six-transistor (6T) bit cell suffers from low noise margins, so 8T-, 9T- and 10T-bit cells have been proposed to increase noise margins and enable lower supply voltages than are possible with the 6T cell. Because 8T- or 9T-bit cells can operate at lower voltages, the increased leakage energy due to higher transistor counts is offset by the decrease in leakage per transistor. Increased noise margins can enable higher threshold transistors to be used to reduce leakage [26]. 9T and 10T cells have achieved very low leakage levels at very low voltages [27,28]. The potential for implementing these design innovations further increases our confidence that leakage energy will not come to dominate power consumption to such an extent that the increased leakage of a scratchpad memory overwhelms its dynamic power savings for the foreseeable future.

6. WIMS Microcontroller

The WIMS microcontroller has been implemented in both 180- and 65-nm processes [2,29], as depicted in Figures 6 and 7, with a small scratchpad memory to augment the on-chip main memory banks. Because it has had relatively few architectural changes between generations, it is a good test bed for evaluating the advantages of scratchpad memories as processes are scaled. Measured data from both implementations will be used to investigate the implications of process differences on the energy savings available from small scratchpad memories, and Intel's transistor data will be used to project energy savings for a hypothetical 32-nm version of the microcontroller.

The WIMS microcontroller is designed for low-power embedded applications, such as environment and health monitoring. Its scratchpad memory is 1 KB. This memory is entirely software-controlled, but is managed in much the same way as a loop cache by the WIMS compiler. The microcontroller has a three-stage pipeline and runs at more than 200 MHz; the maximum frequency is unknown due to tester limitations. Its on-chip main memory is 32 KB, composed of four 8-KB banks.

Although the scratchpad memory in the WIMS microcontroller was originally referred to as a loop cache [10], it will be referred to as a scratchpad memory here to be consistent with conventional terminology [9], because it does not have a hardware controller or tag comparison circuitry. Unlike a typical cache or a hardware-controlled loop cache, the scratchpad is memory-mapped, so that it can be managed by the compiler.

Figure 6. A die photo of the 180-nm WIMS microcontroller.

Figure 7. The layout of the third-generation WIMS microcontroller. The die photo on top shows metallization with block outlines added. The lower image is a screenshot of the digital logic layout.

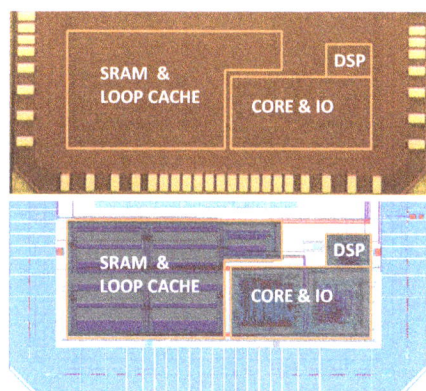

6.1. Measured Data

Measurements from the 65-nm WIMS microcontroller were recorded at the University of Utah using the Verigy 93000 PS400 SOC tester. The memories are in a separate power domain, so that their power consumption could be easily separated from that of the core. To estimate energy savings, three variables were measured: main memory fetch energy (MFE), scratchpad fetch energy (SFE) and leakage energy (LE). Scratchpad leakage energy (SLE) was estimated by using memory compiler reports for a commercial 65-nm memory compiler. The memory compiler reports estimate the leakage power of the memory and showed that SLE is approximately 3% of LE, while the main memory leakage energy (MLE) is 97% of LE. This result is intuitive, because the scratchpad memory is approximately 3% of the system memory. To measure MFE and SFE, a loop of code was executed from main memory and subsequently from scratchpad memory, and memory power was measured while running at various frequencies and voltages. The static memory power draw was also measured for each voltage. The power draw was divided by the clock cycle to find the energy consumed per clock cycle. MFE and SFE were found by subtracting LE from the energy measured during execution from the main memory and scratchpad memory, respectively.

Energy savings were then calculated using (2), but simplified by setting AR to one, because the entire program was placed in the scratchpad memory:

$$1 - ES = \frac{SFE + MLE + SLE}{MFE + MLE} \tag{4}$$

The percentage of energy saved using the scratchpad memory *versus* main memory in the 65-nm WIMS microcontroller implementation is depicted in Figure 8 for voltages between 0.9 and 1.2 V and frequencies between 80 and 200 MHz. As explained above, the total energy includes leakage energy. The scratchpad memory saves 10%–25% of total energy across all measurements, with the maximum savings achieved at the lowest voltages. These large total energy savings justify the use of a scratchpad memory in any application that has frequently used code or data and where dynamic power is a major contributor to total power.

Figure 8. Measured memory energy savings from executing code from the scratchpad memory instead of the main memory on the 65-nm WIMS microcontroller.

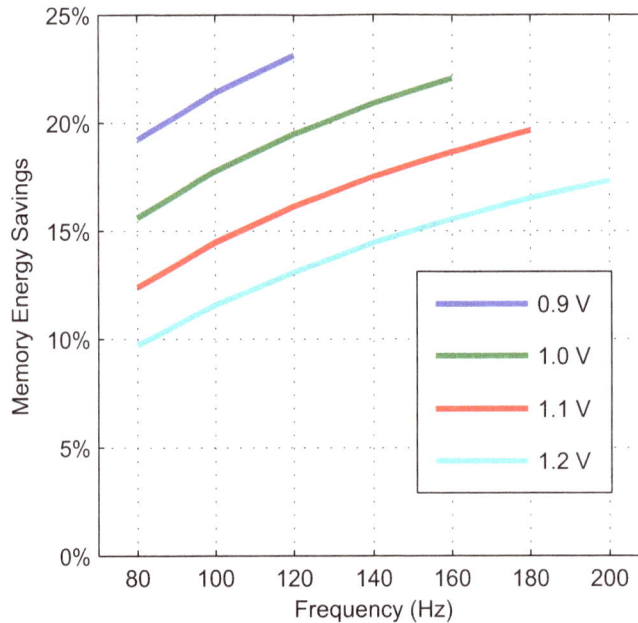

Even in efficient operating regimes, using the lowest possible supply voltage for a given clock frequency, leakage is responsible for approximately 45% of the total energy for the 65-nm WIMS microcontroller. This relatively high leakage ratio limits the total energy savings, because the scratchpad memory can only reduce dynamic energy consumption.

6.2. 32-nm Projections

To investigate whether recent scaling trends are advantageous or detrimental to scratchpad memories, the energy savings of a hypothetical 32-nm version of the WIMS microcontroller are projected. The 32-nm process was selected instead of the 22-nm process, because the 32-nm process is planar, so the assumptions necessary for the projection are more justifiable. To project the values of SFE, MFE, SLE and MLE that would likely be attained by a 32-nm version of the WIMS microcontroller, the energy measurements from the 65-nm version are scaled according to the transistor trends depicted in Figure 4. First, transistor parameters must be related to the energy variables in (4). Leakage energy per clock period (LE) is proportional to the supply voltage (V) multiplied by the leakage current (I_{leak}) and the clock period (t_f):

$$LE \propto V \cdot I_{leak} \cdot t_f \tag{5}$$

Leakage current can be further broken down into leakage per micron of gate width ($I_{leak}/\mu m$) multiplied by the average gate width (G_W):

$$LE \propto V \cdot \frac{I_{leak}}{\mu m} \cdot G_W \cdot t_f \tag{6}$$

Because the scratchpad memory primarily saves active energy, at the expense of a small increase in leakage energy, projected results are conservative if the decrease in leakage is underestimated and the decrease in active energy is overestimated. Based on the data presented in Figure 4, $I_{leak}/\mu m$ in 32 nm is 30% of its value in the 65-nm process, while the voltage is reduced from 1.2 V to 1.0 V. To keep results conservative, no reduction in clock period was assumed. Therefore, in a 32-nm process, LE/G_W, relative to the 65-nm process, is approximately:

$$\frac{LE_{32}/G_{W32}}{LE_{65}/G_{W65}} = \frac{I_{leak32}/\mu m}{I_{leak65}/\mu m} \cdot \frac{V_{32}}{V_{65}} \tag{7}$$
$$= 0.3 \cdot \frac{1.0 \text{ V}}{1.2 \text{ V}} = 0.25$$

where the "32" and "65" subscripts indicate whether the parameters are for 32 nm or 65 nm, respectively. To estimate G_{W32}/G_{W65}, it is assumed that the reduction in gate area (A_{32}/A_{65}) is approximately the same as the reduction in bit cell area. For the lowest power option, minimum G_L is 55 nm in Intel's 65-nm technology and 46 nm in Intel's 32-nm technology. The bit cell area decreased from 0.68 μm^2 in 65 nm to 0.171 μm^2 in 32 nm (to 25% of its original value) [19,21]. Therefore,

$$\frac{G_{W32}}{G_{W65}} = \frac{A_{32}}{A_{65}} \cdot \frac{G_{L65}}{G_{L32}} \tag{8}$$
$$= \frac{0.171 \ \mu m^2}{0.68 \ \mu m^2} \cdot \frac{55 \text{ nm}}{46 \text{ nm}} = 0.3$$

In sum, LE in 32 nm, relative to 65 nm, is estimated to be:

$$\frac{LE_{32}}{LE_{65}} = \frac{LE_{32}/G_{W32}}{LE_{65}/G_{W65}} \cdot \frac{G_{W32}}{G_{W65}} \tag{9}$$
$$= 0.25 \cdot 0.3 = 0.075$$

Now that the leakage scaling ratio has been found, the fetch energy scaling ratio will be considered. The fetch energy (FE) is assumed to be proportional to dynamic switching energy, which is proportional to the switching capacitance of the wires and transistors (C) times the supply voltage (V) squared:

$$FE \propto C \cdot V^2 = (C_t + C_w) \cdot V^2 \tag{10}$$

where C_t represents capacitance due to transistors and C_w represents the capacitance due to wires. First, the division between wire capacitance and transistor capacitance should be estimated for our

SRAM. This division was studied for wires of different lengths in a 130-nm process [30]. This study can be used to indicate the division between C_w and C_t if the WIMS microcontroller SRAM's wire lengths can be estimated and then converted to the 130-nm equivalent. The 65-nm WIMS SRAM uses 8-KB banks, each of which are 320 μm × 170 μm. Its wires are therefore up to 320 μm, and these lengths should be doubled to up to 640 μm to be equivalent to a 130-nm process. Magen *et al.* [30] indicates that for wires up to 640 μm long, C_w is approximately 40% of total C.

C_t can be estimated by assuming that C_t is roughly proportional to the gate area of the transistor and that the decrease in the gate area of the transistor is proportional to the decrease in the area of one SRAM bit cell. This last assumption was also made by (8). The decrease in the SRAM bit cell area is from 0.68 μm^2 in 65 nm to 0.171 μm^2 in 32 nm. Therefore,

$$\frac{C_{t32}}{C_{t65}} = \frac{0.171 \ \mu m^2}{0.68 \ \mu m^2} = 0.25 \tag{11}$$

To estimate C_{w32}/C_{w65}, the predictive technology model was used to find the wire capacitance in the 65-nm process [31], and 2011 ITRS parameters were used to estimate the wire capacitance in the 32-nm process [32]. Using the typical wire dimensions suggested by the predictive technology model, a typical wire in 65-nm has about 1.46 pF/cm of capacitance. Table INTC2 in the Interconnect section of the 2011 ITRS lists intermediate wires having 1.8 pF/cm of capacitance in 2011, the year the 32-nm node was introduced. This is a slight increase, but the length of each wire is halved, so the total change in C_w is:

$$\frac{C_{w32}}{C_{w65}} = 0.5 \cdot \frac{1.8 \frac{pF}{cm}}{1.46 \frac{pF}{cm}} = 0.62 \tag{12}$$

Although the scaling of the wire capacitance is worse than transistor capacitance scaling, this poor scaling actually makes scratchpad memories more attractive, because it increases dynamic power (which scratchpad memories reduce) relative to leakage power. Now, it is possible to find the overall capacitance scaling:

$$\begin{aligned} \frac{C_{32}}{C_{65}} &= \frac{C_{t32} + C_{w32}}{C_{65}} \\ &= \frac{C_{t32}}{C_{65}} + \frac{C_{w32}}{C_{65}} \\ &= \frac{C_{t65}}{C_{65}} \cdot \frac{C_{t32}}{C_{t65}} + \frac{C_{w65}}{C_{65}} \cdot \frac{C_{w32}}{C_{w65}} \\ &= 0.6 \cdot 0.25 + 0.4 \cdot 0.62 = 0.398 \end{aligned} \tag{13}$$

FE_{32}/FE_{65} can be calculated using (10):

$$\begin{aligned} \frac{FE_{32}}{FE_{65}} &= \frac{C_{32}}{C_{65}} \cdot \left(\frac{V_{32}}{V_{65}}\right)^2 \\ &= 0.398 \cdot \left(\frac{1.0 \ V}{1.2 \ V}\right)^2 = 0.276 \end{aligned} \tag{14}$$

Using these scaling figures for LE and FE, the energy savings of the WIMS scratchpad memory will be projected for 32 nm, based on the 65-nm measured data.

6.3. WIMS Scratchpad Memory Approach

The WIMS scratchpad memory is compiler-controlled. When use of the scratchpad memory was analyzed [10], two compiler management methods were implemented: "static" and "dynamic." The static method identifies the most frequently used loops and data objects and places as many as fit into the scratchpad memory for the duration of the program execution. The dynamic method uses a trace-selection algorithm to determine whether the energy savings of having a certain program loop in the scratchpad memory is worth the energy of copying it into the scratchpad. Copy instructions are then inserted in the program to ensure that the selected loop or data is copied to the scratchpad memory before it is executed or accessed.

The example program depicted in Figure 9 will be used to illustrate the differences between the static and dynamic methods, after each algorithm is summarized. The compiler routine for the static allocation algorithm can be summarized as follows:

1. Identify the most executed blocks of code.
2. Assign these blocks of code, in the order of the most used first, to the scratchpad memory, until it is full.
3. Redirect instruction fetch (when entering code block), subroutine calls and references to these blocks to the new locations in the scratchpad memory.

Figure 9. An example program in which each numbered block represents a piece of code and an arrow represents a possible execution path. The boxes shaded the darkest are the most frequently executed. The static allocation method will hold Blocks 4 and 5 in the scratchpad memory for the duration of the execution of the program, whereas the dynamic allocation method will replace those blocks with Blocks 10 and 11 prior to their execution.

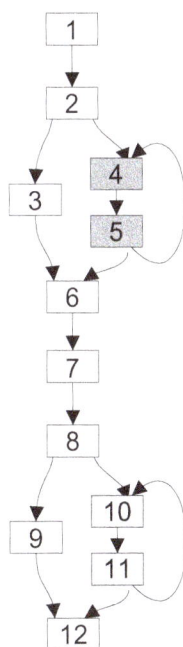

In the example program in Figure 9, the loop consisting of Blocks 4 and 5 is the most frequently executed code in the program, followed by the loop consisting of Blocks 10 and 11. If only two blocks can fit in the scratchpad memory, then the static method will place Blocks 4 and 5 in the scratchpad memory for the duration of the execution of the program.

In the dynamic method, the compiler uses profiling to determine how much energy would be saved by moving a block to the scratchpad memory. If the compiler determines that the energy saved by running a block from the scratchpad memory is worth the energy of copying the code to the scratchpad memory, then it will assign those code blocks to addresses in the scratchpad memory. The dynamic allocation algorithm can be summarized as:

1. Identify blocks of code, such that the energy to copy the code to the scratchpad memory is less than the energy saved by executing the code from the scratchpad memory.
2. Assign these blocks of code addresses in the scratchpad memory and redirect references to them to their new addresses.
3. Insert copy instructions immediately prior to each of these blocks, so that the block will be in the scratchpad memory prior to its execution or prior to its being called if it is a subroutine.
4. Consolidate copy instructions to avoid unnecessary copying.

In the example program in Figure 9, if the compiler determines that the energy savings of running Blocks 4, 5, 10 and 11 is worth copying them to the scratchpad memory, then each block will be placed in the scratchpad memory and instructions will be inserted to copy each block to the scratchpad memory prior to its execution. Assuming two blocks can fit in the scratchpad memory, the copy instructions for Blocks 4 and 5 will be consolidated, as well as the copy instructions for Blocks 10 and 11.

The compiler uses execution profile information and an energy benefit heuristic to determine which blocks to move to the scratchpad memory. Details of the compilation procedures used in the static and dynamic allocation methods are beyond the scope of this paper, but are available in [10].

The energy savings of the dynamic and static methods will be presented for the 180- and 65-nm processes based on measured data, and the energy savings of a hypothetical 32-nm process will be projected.

6.4. Static Benchmarks

The benchmarks used to compare the scratchpad memory energy savings in the 180-, 65- and 32-nm processes are the same as those used to compare the static and dynamic allocation methods in [10]. Those benchmarks are a subset of the MediaBench [33] and MiBench [34] benchmarks, oriented towards embedded systems. This selection of benchmarks is convenient, because scratchpad memory access rates for each of these benchmarks have already been published in [10] for the WIMS microcontroller, using the static allocation method. This data can be used to project the energy savings (ES) of the microcontroller using the model represented by (2), repeated here with a slight modification for convenience:

$$1 - ES = \frac{AR \cdot SFE + (1 - AR) \cdot MFE + LE}{MFE + MLE} \tag{15}$$

Leakage energy (LE) has replaced $MLE + SLE$ for brevity ($LE = MLE + SLE$). The access rates (AR) are found in [10]. SFE, MFE and LE were measured for 0.9 V and 120 MHz, as described in Section 6.1, and $MLE = 0.97 \cdot LE$, as noted in the same section. The frequency of 120 MHz was selected, because it was the highest operational frequency for the microcontroller with a 0.9-V supply voltage. Projected energy savings are presented in Figure 10 for a 256 B scratchpad memory. The 65-nm WIMS microcontroller has a larger 1 KB scratchpad memory, but scratchpad access rates were not available for all benchmarks for that larger size, so the results presented are conservative. The 32-nm projections of each of these variables were made using the method described in Section 6.2.

Figure 10. Comparison of energy savings from a scratchpad memory in 180-, 65- and 32-nm technologies, using the static allocation method.

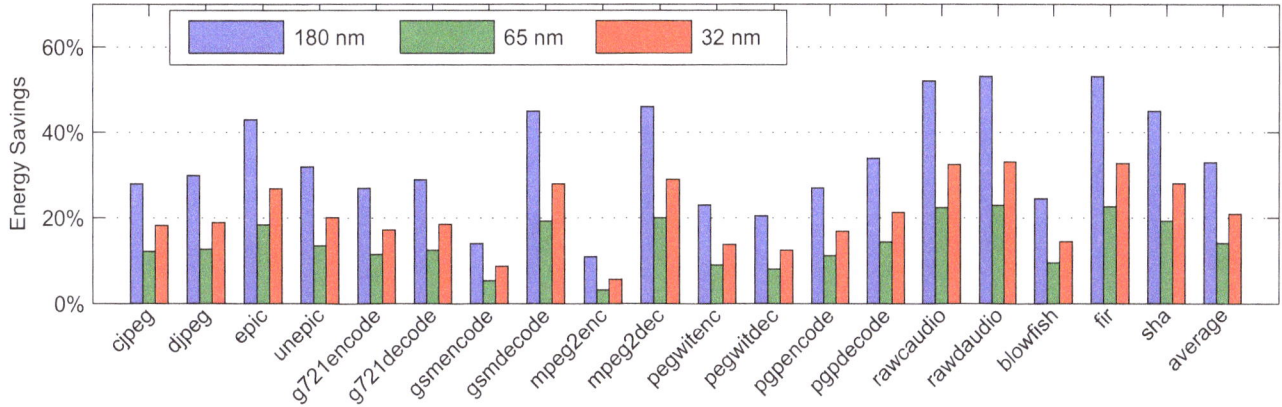

The scratchpad memory is limited to saving dynamic power, which is only ~55% of the WIMS microcontroller's total memory power. The remaining ~45% is due to leakage. Thus, despite savings as high as 40% in terms of dynamic energy consumption, Figure 10 shows lower total energy savings because of the high leakage power.

6.5. Dynamic Benchmarks

To use the measured data to project energy savings from using the dynamic allocation scheme, two new variables are added to the model:

$$1 - ES = \frac{AR \cdot SFE + (1 - AR) \cdot MFE + CO \cdot (MFE + SWE) + LE}{MFE + MLE} \qquad (16)$$

CO (copy overhead ratio) represents the impact of the additional instructions necessary to copy code from the main memory to the scratchpad memory during execution, per memory access. Because a copy involves a read from main memory and a write to scratchpad memory, CO is proportional to MFE and SWE (scratchpad write energy). For the 180-nm processor, all of these variables are reported for each benchmark except for CO, which was solved for using Equation (16).

For the 65-nm processor and 32-nm projection, the copy overhead ratio CO is the same as it is in the 180-nm processor, because the memory sizes and ISA (instruction set architecture) are the same, so the number of copy instructions executed to fill the scratchpad during execution will be the same. SFE was substituted for SWE, because the available SFE data were more reliable. Tests on the 180-nm processor indicated that the read and write energies are similar; specifically, the read and write energies of a 512 B memory were virtually identical, and the read and write energies of a 1 KB memory (the size of the scratchpad on the WIMS microcontroller) were within 20% of each other. ES was then calculated for the 65-nm processor using (16). The 32-nm projections were again made using the method described in Section 6.2.

The results depicted in Figures 10 and 11 reveal that: (1) the dynamic allocation method generally increases the energy savings achieved by the scratchpad memory; (2) the scratchpad memory saves substantial energy in every examined process; and (3) that energy savings expected in a 32-nm version of the microcontroller should be even greater than observed in the 65-nm version.

Figure 11. Comparison of energy savings from a scratchpad memory in 180-, 65- and 32-nm technologies, using the dynamic allocation method.

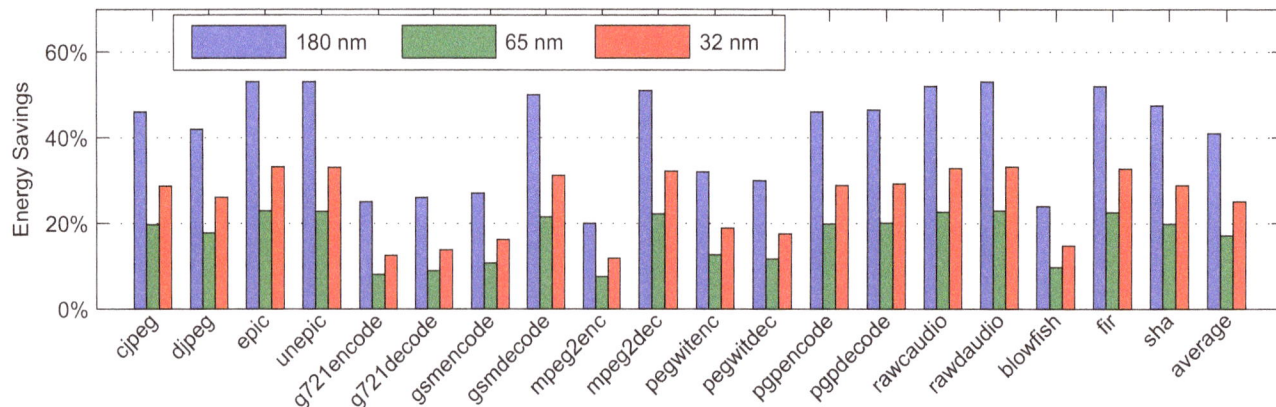

7. Conclusions

Despite the high leakage ratio at the 65-nm process node, this study has shown that the energy saved by using scratchpad memories continues to merit their consideration. In nodes with lower leakage ratios, such as the 45- or 32-nm nodes, or with memories that are optimized to minimize leakage power, scratchpad memories are even more attractive. In particular, the largest gains were found for frequent memory accesses and minimal memory supply voltage. Although this study focused on using scratchpad memories for instruction storage, we believe these conclusions will hold if the scratchpad memories are used for data storage, as well. Applications with low memory-access frequency or that use processes with high leakage-to-dynamic power ratios will see limited savings from scratchpad memories. These conclusions were supported by data measured from the WIMS microcontroller implemented in both 180- and 65-nm processes. CACTI simulations were used to demonstrate how process characteristics impact SRAM energy consumption and scratchpad memory energy savings. Intel process technology reports indicate that process innovations will continue to decrease leakage energy. These data can guide the design process at multiple levels from software to architecture to process selection. In many applications, scratchpad memories will provide significant savings in dynamic power consumption at little cost in leakage power, even in advanced semiconductor processes. This power savings is likely to continue for the process nodes of the foreseeable future.

Acknowledgments

This work was supported by the Engineering Research Center Program of the National Science Foundation under Award No. EEC-9986866. This work made use of University of Utah Shared

Facilities supported, in part, by the MRSEC (Materials Research Science and Engineering Centers) Program of the NSF under Award No. DMR-1121252. Chip testing was enabled by the use of VTRAN (a vector translation program) from the kind support of Source III. Chip fabrication was made possible by MOSIS (Metal Oxide Semiconductor Implementation Service, an organization that provides multi-project wafers). Toren Monson assisted in the testing of the 65-nm WIMS microcontroller.

Author Contributions

Bennion Redd wrote the first drafts of the text and incorporated the other authors' feedback, as well as led the testing of the 65-nm WIMS microcontroller on the Verigy 93000. Spencer Kellis and Nathaniel Gaskin led the tape-out process to prepare the chip design for fabrication at MOSIS, and Richard Brown supported and directed work on the WIMS microcontroller, consulted on design decisions, suggested research ideas and offered valuable feedback on revisions of the paper.

Conflicts of Interest

The authors declare no conflict of interest.

References

1. Redd, B.; Kellis, S.; Gaskin, N.; Brown, R. Scratchpad Memories in the Context of Process Scaling. In Proceedings of the IEEE 54th International Midwest Symposium on the Circuits and Systems (MWSCAS), Seoul, Korea, 7–10 August 2011.

2. Kellis, S.; Gaskin, N.; Redd, B.; Campbell, J.; Brown, R. Energy Profile of a Microcontroller for Neural Prosthetic Application. In Proceedings of the IEEE International Symposium on Circuits and Systems, Paris, France, 30 May–2 June 2010; pp. 3841–3844.

3. Steinke, S.; Wehmeyer, L.; Bo-Sik, L.; Marwedel, P. Assigning Program and Data Objects to Scratchpad for Energy reduction. In Proceedings of the Design, Automation and Test in Europe Conference and Exhibition, Paris, France, 4–8 March 2002; pp. 409–415.

4. Kin, J.; Gupta, M.; Mangione-Smith, W.H. The Filter Cache: An Energy Efficient Memory Structure. In Proceedings of the 30th Annual ACM/IEEE International Symposium Microarchitecture, Washington, DC, USA, 1–3 December 1997; pp. 184–193.

5. Bellas, N.; Hajj, I.; Polychronopoulos, C.; Stamoulis, G. Energy and Performance Improvements in Microprocessor Design Using a Loop Cache. In Proceedings of the International Conference on Computer Design (ICCD '99), Austin, TX, USA, 10–13 October 1999; pp. 378–383.

6. Lea Hwang, L.; Moyer, B.; Arends, J. Instruction Fetch Energy Reduction Using Loop Caches for Embedded Applications with Small Tight Loops. In Proceedings of the Low Power Electronics and Design, San Diego, CA, USA, 17 August 1999; pp. 267–269.

7. Gordon-Ross, A.; Cotterell, S.; Vahid, F. Exploiting fixed programs in embedded systems: A loop cache example. *IEEE Comp. Archit. Lett.* **2002**, *1*, 2.

8. Egger, B.; Lee, J.; Shin, H. Scratchpad Memory Management for Portable Systems with a Memory Management Unit. In Proceedings of the 6th ACM & IEEE International Conference on Embedded Software (EMSOFT '06), Seoul, Korea, 22–25 October 2006; ACM: New York, NY, USA, 2006; pp. 321–330.

9. Verma, M.; Wehmeyer, L.; Marwedel, P. Cache-Aware Scratchpad Allocation Algorithm. In Proceedings of the Design, Automation and Test in Europe Conference and Exhibition, Washington, DC, USA, 16–20 February 2004; pp. 1264–1269.

10. Ravindran, R.A.; Nagarkar, P.D.; Dasika, G.S.; Marsman, E.D.; Senger, R.M.; Mahlke, S.A.; Brown, R.B. Compiler Managed Dynamic Instruction Placement in a Low-Power Code cache. In Proceedings of the International Symposium on Code Generation and Optimization (CGO), Washington, DC, USA, 20–23 March 2005; pp. 179–190.

11. Borkar, S. Design challenges of technology scaling. *Micro IEEE* **1999**, *19*, 23–29.

12. Guangyu, C.; Feihui, L.; Ozturk, O.; Guilin, C.; Kandemir, M.; Kolcu, I. Leakage-Aware SPM Management. In Proceedings of the IEEE Computer Society Annual Symposium on Emerging VLSI Technologies and Architectures, Karlsruhe, Germany, 2–3 March 2006.

13. Kandemir, M.; Irwin, M.J.; Chen, G.; Kolcu, I. Compiler-guided leakage optimization for banked scratch-pad memories. *IEEE Trans. Larg. Scale Integr. (VLSI) Syst.* **2005**, *13*, 1136–1146.

14. Huangfu, Y.; Zhang, W. Compiler-based approach to reducing leakage energy of instruction scratch-pad memories. In Proceedings of the 2013 IEEE 31st International Conference on Computer Design (ICCD), Asheville, NC, USA, 6–9 October 2013; pp. 439–442.

15. Takase, H.; Tomiyama, H.; Zeng, G.; Takada, H. Energy efficiency of scratch-pad memory in deep submicron domains: An empirical study. *IEICE Electron. Express* **2008**, *5*, 1010–1016.

16. Thoziyoor, S.; Muralimanohar, N.; Ahn, J.; Jouppi, N. CACTI 5.1. Available online: http://www.hpl.hp.com/techreports/2008/HPL-2008-20.html (accessed on 5 June 2014).

17. 2005 International Technology Roadmap for Semiconductors. Available online: http://www.itrs.net/reports.html (accessed on 5 June 2014).

18. Wilton, S.; Jouppi, N. CACTI: An enhanced cache access and cycle time model. *IEEE J. Solid-State Circuits* **1996**, *31*, 677–688.

19. Jan, C.H.; Bai, P.; Choi, J.; Curello, G.; Jacobs, S.; Jeong, J.; Johnson, K.; Jones, D.; Klopcic, S.; Lin, J.; *et al.* A 65 nm Ultra Low Power Logic Platform Technology Using Uni-Axially Strained-Silicon Transistors. In Proceedings of the 2005 IEEE International Electron Devices Meeting (IEDM) Technical Digest, Washington, DC, USA, 5 December 2005; pp. 60–63.

20. Jan, C.H.; Bai, P.; Biswas, S.; Buehler, M.; Chen, Z.P.; Curello, G.; Gannavaram, S.; Hafez, W.; He, J.; Hicks, J.; *et al.* A 45 nm Low Power System-on-Chip Technology with Dual Gate (Logic and I/O) High-k/Metal Gate Strained Silicon Transistors. In Proceedings of the 2008 IEEE International Electron Devices Meeting (IEDM 2008), San Francisco, CA, USA, 15–17 December 2008.

21. Jan, C.H.; Agostinelli, M.; Buehler, M.; Chen, Z.P.; Choi, S.J.; Curello, G.; Deshpande, H.; Gannavaram, S.; Hafez, W.; Jalan, U.; *et al.* A 32 nm SoC Platform Technology with 2nd Generation High-k/Metal Gate Transistors Optimized for Ultra Low Power, High Performance, and High Density Product Applications. In Proceedings of the 2009 IEEE International Electron Devices Meeting (IEDM), Baltimore, MD, USA, 7–9 December 2009.

22. Jan, C.H.; Bhattacharya, U.; Brain, R.; Choi, S.J.; Curello, G.; Gupta, G.; Hafez, W.; Jang, M.; Kang, M.; Komeyli, K.; *et al.* A 22 nm SoC Platform Technology Featuring 3-D Tri-gate and High-k/metal gate, optimized for ultra low power, high performance and high density SoC applications. In Proceedings of the 2012 IEEE International Electron Devices Meeting (IEDM), San Francisco, CA, USA, 10–13 December 2012; pp. 3.1.1 – 3.1.4.

23. Dennard, R.; Gaensslen, F.; Rideout, V.; Bassous, E.; LeBlanc, A. Design of ion-implanted MOSFET's with very small physical dimensions. *Solid-State Circuits IEEE J.* **1974**, *9*, 256–268.

24. Yang, H.S.; Wong, R.; Hasumi, R.; Gao, Y.; Kim, N.S.; Lee, D.H.; Badrudduza, S.; Nair, D.; Ostermayr, M.; Kang, H.; *et al.* Scaling of 32 nm Low Power SRAM with High-K Metal Gate. In Proceedings of the 2008 IEEE International Electron Devices Meeting (IEDM), San Francisco, CA, USA, 15–17 December 2008.

25. Fujita, K.; Torii, Y.; Hori, M.; Oh, J.; Shifren, L.; Ranade, P.; Nakagawa, M.; Okabe, K.; Miyake, T.; Ohkoshi, K.; *et al.* Advanced Channel Engineering Achieving Aggressive Reduction of VT Variation for Ultra-Low-Power Applications. In Proceedings of the 2011 IEEE International Electron Devices Meeting (IEDM), Washington, DC, USA, 5–7 December 2011; pp. 32.3.1–32.3.4.

26. Athe, P.; Dasgupta, S. A Comparative Study of 6T, 8T and 9T Decanano SRAM Cell. In Proceedings of the 2009 IEEE International Symposium on Industrial Electronics Applications (ISIEA 2009), Kuala Lumpur, Malaysia, 4–6 October 2009; Volume 2, pp. 889–894.

27. Calhoun, B.H.; Chandrakasan, A. A 256kb Sub-threshold SRAM in 65 nm CMOS. In Proceedings of the 2006 IEEE International Solid-State Circuits Conference Digest of Technical Papers (ISSCC 2006), San Francisco, CA, USA, 6–9 February 2006; pp. 2592–2601.

28. Lin, S.; Kim, Y.B.; Lombardi, F. A Low Leakage 9T Sram Cell for Ultra-Low Power Operation. In Proceedings of the 18th ACM Great Lakes Symposium on VLSI, Orlando, FL, USA, 4–6 May 2008; ACM Press: New York, NY, USA, 2008; pp. 123–126.

29. Marsman, E.; Senger, R.; McCorquodale, M. A 16-bit Low-Power Microcontroller with Monolithic MEMS-LC Clocking. In Proceedings of the 2005 IEEE International Symposium on Circuits and Systems (ISCAS 2005), Kobe, Japan, 2005; pp. 624–627.

30. Magen, N.; Kolodny, A.; Weiser, U.; Shamir, N. Interconnect-Power Dissipation in a Microprocessor. In Proceedings of the 2004 International Workshop on System Level Interconnect Prediction (SLIP '04), Paris, France, 14–15 February 2004; ACM: New York, NY, USA, 2004; pp. 7–13.

31. Predictive Technology Model. Available online: http://ptm.asu.edu/interconnect.html (accessed on 5 June 2014).

32. 2011 International Technology Roadmap for Semiconductors. Available online: http://www.itrs.net/reports.html (accessed on 5 June 2014).

33. Lee, C.; Potkonjak, M.; Mangione-Smith, W. MediaBench: A Tool for Evaluating and Synthesizing Multimedia and Communications Systems. In Proceedings of the 30th Annual IEEE/ACM International Symposium on Microarchitecture, Research Triangle Park, NC, USA, 1–3 December 1997; pp. 330–335.

34. Guthaus, M.; Ringenberg, J.; Ernst, D.; Austin, T.; Mudge, T.; Brown, R. MiBench: A Free, Commercially Representative Embedded Benchmark Suite. In Proceedings of the 2001 IEEE International Workshop on Workload Characterization (WWC-4 2001), Washington, DC, USA, 2 December 2001; pp. 3–14.

9

39 fJ/bit On-Chip Identification of Wireless Sensors Based on Manufacturing Variation

Jonathan F. Bolus *, Benton H. Calhoun and Travis N. Blalock

Department of Electrical & Computer Engineering, University of Virginia, 351 McCormick Rd., Charlottesville, VA 22904, USA; E-mails: bcalhoun@virginia.edu (B.H.C.); tblalock@virginia.edu (T.N.B.)

* Author to whom correspondence should be addressed; E-Mail: jfbolus@virginia.edu

Abstract: A 39 fJ/bit IC identification system based on FET mismatch is presented and implemented in a 130 nm CMOS process. ID bits are generated based on the ΔV_T between identically drawn NMOS devices due to manufacturing variation, and the ID cell structure allows for the characterization of ID bit reliability by characterizing ΔV_T. An addressing scheme is also presented that allows for reliable on-chip identification of ICs in the presence of unreliable ID bits. An example implementation is presented that can address 1000 unique ICs, composed of 31 ID bits and having an error rate less than 10^{-6}, with up to 21 unreliable bits.

Keywords: chip identification; low-power electronics; radio-frequency identification; wireless sensor networks; PUF

1. Introduction

Recent advances in the design of wirelessly powered, millimeter-scale sensor tags will allow for the construction of increasingly small sensors for a variety of applications [1–5]. In the case of wirelessly powered sensors like RFID tags, however, decreasing size leads to decreased power delivery due to reduced antenna area. All sensor components must therefore be designed for minimum power consumption. A common component of such sensors is a non-volatile memory used to store a unique identification (ID) number. However, for very small sensors, implementation of such a memory is non-trivial because of the low supply voltage and available energy, which can be insufficient to program a conventional non-volatile (NV) memory such as Flash or EEPROM. An alternative would be to program each sensor at its time of manufacture, for example by a physical electrical connection that could supply the energy required for a NV memory, or by an array of laser-blown fuses. These approaches require extra masks or post-manufacturing steps, which raise the individual cost of the supposedly low-cost devices.

An alternative is an identification system based on the individual variation of ICs due to the CMOS manufacturing process [6]. A suitably designed circuit could be sensitive to these variations, and produce a string of bits that is random between chips, but temporally static, to be used as an ID number. This could be thought of as a particular type of NV memory that is "programmed" once at the time of manufacture with random data. Such variation sensitive circuits also have security and cryptographic applications, where they can be used to form Physical Unclonable Functions (PUFs) that allow for authentication and secret key generation [7,8].

In this paper, we present a circuit that generates random identification numbers based on manufacturing variation, with lower energy than reported in prior published work. We also present an addressing protocol that allows for reliable on-chip identification in the presence of unreliable bits. This reduces the amount of data that must be transmitted off-chip, lowering the energy consumption of the sensor. An analysis of the reliability of the system is also presented. While intended for the low-energy identification of wireless sensors, this same circuit could be used as part of a low-energy secret key generator for cryptographic applications [9].

2. System Overview and Design

The central element of the identification system is the ID generator, a circuit that produces an N-bit ID number based on small variations in the IC due to manufacturing. The number should be random, with each bit having equal likelihood of being "0" or "1". The random event is the manufacture of the IC, which occurs only once.

Many circuits exhibit measurable differences due to manufacturing variation that could be used to generate random, unique data. These include memories, such as SRAM power-up state [10] and static noise margin [11], and DRAM retention fails [12]. Variations in delay lines and arbiters [13], ring-oscillators [9], scan-chain power-up state [14], and cross-coupled inverters [15–17], as well as direct measurements of the drain current of individual FETs [6] have also been examined for the generation of unique identification data and PUFs.

All random ID implementations have the property that not all of the random ID bits are necessarily reliable. The ID generator circuit should produce the same ID number every time it is activated, but in such systems some bits are more reliable than others, and unreliable bits may occasionally flip between successive ID generation events. Ultimately, this could cause errors in the identification process. Current literature solves this problem by requiring each sensor to transmit its generated ID number back to the external reader, which increases the amount of data that must be transmitted. However, for such systems, wireless communication is frequently the largest energy consumer, so any reduction in transmitted data directly improves system performance. It has been observed that prior knowledge of which bits are unreliable can be used to increase the accuracy of identification [15,18], and that the reliability of individual bits can be determined by examining the magnitude of the underlying variation [17].

Ideally, the ID generator bits would not be temporally random: successive reads or power-on cycles would always produce the same ID number for the lifetime of the IC. However, due to the nature of manufacturing variations and the presence of electrical noise, the value of individual bits may vary over multiple read operations. We can model the output of the ID generator, which we will call the generated code, G, as a discrete, N-bit random variable. For each bit, G_i we can assign a probability, p, such that the probability of bit G_i evaluating to "1" is p, and evaluating to "0" is $1 - p$. An N-bit ID generator can then be completely described by a sequence of N probabilities $(p_0, p_1, ..., p_{N-1})$.

Based on this ID generator, each IC can then be assigned an ID code C, where

$$C_i = \begin{cases} 1 & p_i > 1/2 \\ 0 & p_i \leq 1/2 \end{cases} \tag{1}$$

In other words, the chip code C is the most likely value of the random variable G, the output of one read of the ID generator. It is this number that will be used to identify the IC.

2.1. Remote Identification

When an external reader attempts to locate a particular IC among a population of ICs, variations in the output of the ID generators may cause errors. For example, consider an external reader that tries to locate IC A by transmitting the chip code for chip A, C_A. All chips in the population receive this code, and compare it to their generated codes, G. If $C_A = G_A$ then chip A concludes it is being addressed and can reply to the external reader. However, this may lead to false negative errors if unreliable bits in chip A's ID generator result in $C_A \neq G_A$.

The common way to deal with this problem is to use a slightly different approach. To find chip A in a population of M chips, an external reader transmits an inquiry to all the chips, which all reply with their generated codes, $(G_0, G_1, ..., G_{M-1})$. The external reader then computes the Hamming distance between all the received codes and chip code A, $H(C_A, G_i)$ for all i. The value of G_i that has the smallest Hamming distance to C_A is then concluded to have come from chip A.

We call this approach remote identification, since the target chip A is unable to positively conclude it is being addressed. Rather, identification happens remotely (from the perspective of the chip), at the external reader. The primary drawback of this approach is that the chip must transmit its generated code, increasing the energy consumed by the radio. This may also cause problems for systems involving large numbers of chips: some system for staggering their replies must be implemented to avoid collisions.

2.2. On-Chip Identification

By on-chip identification, we mean an addressing protocol by which chips are able to positively conclude they are being addressed. This can be accomplished with knowledge of the reliability of individual ID generator bits. Each bit of the ID generator can be classified as reliable or unreliable, based on each bit's value of p. Although p is a continuous random variable, we can approximate it as a discrete random variable p', where

$$p' = \begin{cases} 1 - \epsilon & 1 - \epsilon \leq p \leq 1 \\ 1/2 & \epsilon < p < 1 - \epsilon \\ \epsilon & 0 \leq p \leq \epsilon \end{cases} \tag{2}$$

The variable ϵ is the threshold of reliability: reliable bits have a probability ϵ of flipping during read, and unreliable bits are treated as evaluating to either "0" or "1" with equal probability.

3. Circuit Implementation

Although there are many types of manufacturing variation that could serve as sources of entropy for a random ID generator, the type chosen for this application is the variation between the threshold voltage, V_T, of two identically drawn FETs. Because the variance of V_T is inversely proportional to the device area [19], both devices are drawn as the minimum size available in the technology. A schematic of an N-bit random ID generator is shown in Figure 1.

Figure 1. N-bit ID generator. The organization is similar to that of a memory, where multiple ID cells share a common set of bit-lines and a sense amplifier.

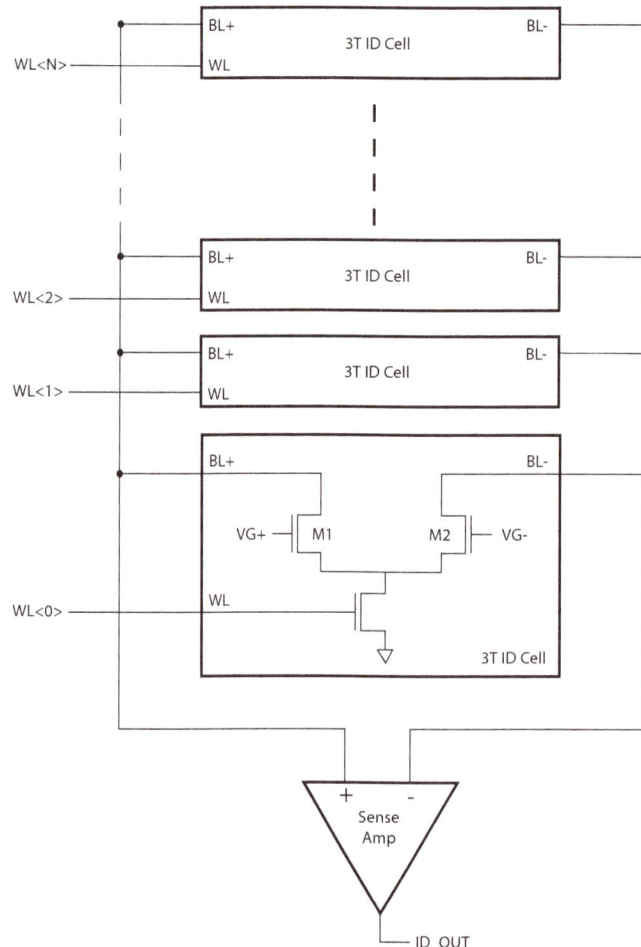

The organization of the ID generator is similar to that of a memory. Each ID cell produces one random bit, based on the V_T difference between M1 and M2. When equal gate voltages are applied to M1 and M2, the V_T difference creates a difference in drain currents, ΔI_D, the polarity of which is detected by the sense amplifier (SA). A column of N ID cells is read sequentially to produce an N-bit random ID number. A simple shift register is used to drive the word-line signals rather than a row decoder, since random access is not necessary, and this requires less energy and area.

Because the SA should be constructed with negligible input referred offset, a single SA built from large devices is constructed, and shared by the column. A schematic of the latch-based sense amplifier is shown in Figure 2. This structure is preferred for low-energy operation since it draws no static current after it has settled to its final state. Sharing the SA among multiple bits also amortizes the SA leakage current.

Figure 2. Schematic of the latch-based sense amplifier in Figure 1.

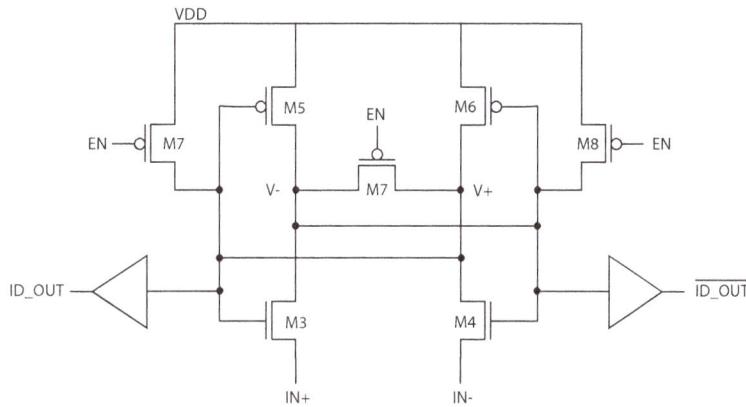

3.1. Noise Analysis

In the absence of electrical noise, reading from a single ID cell would be entirely deterministic. However, when the sense amplifier is activated, the total difference between the ID cell currents is the combination of the inherent offset due to device mismatch and any electrical noise, such that

$$\Delta I = \Delta I_m + \Delta I_n \tag{3}$$

The internal nodes of the sense amplifier, V_+ and V_-, are pre-charged to VDD prior to sensing, and devices M5 and M6 are off at the start of a read operation. When the sense amplifier is activated, the internal nodes are discharged by the ID cell current. When one of the nodes discharges sufficiently to turn on M5 or M6, the positive feedback is engaged and the amplifier settles to a stable state. A simulation of this is shown in Figure 3, showing the three stages of operation.

The total difference voltage developed on the internal SA nodes, $\Delta V = V_+ - V_-$, is the sum of the individual difference voltages due to mismatch and noise.

$$\Delta V = \Delta V_m + \Delta V_n \tag{4}$$

Figure 3. Simulated operation of SA. Stage I: pre-charge, Stage II: integration, Stage III: positive feedback.

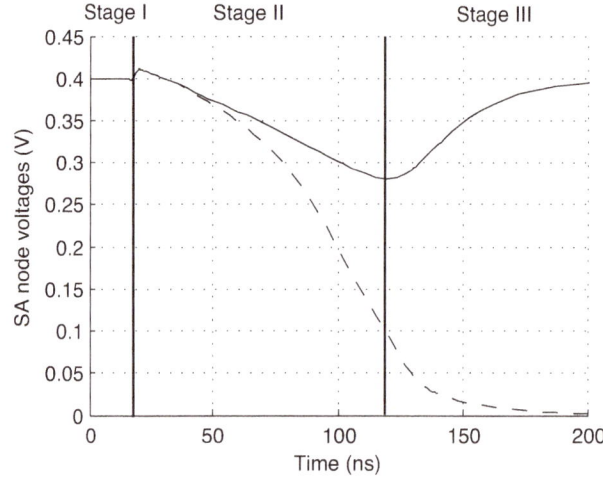

The effect of the mismatch and noise on the circuit output can be calculated. The probability of ΔV being positive at the time the SA feedback is activated is:

$$0 < \Delta V_m + \Delta V_n \tag{5}$$

This is given by:

$$p = P\left(\Delta V_n > -\Delta V_m\right) \tag{6}$$

$$= \int_{-\Delta V_m}^{\infty} \Phi_{\Delta V_n}(v)dv \tag{7}$$

$$= \frac{1}{2}\left[1 + \mathrm{erf}\left(\frac{\Delta V_m}{\sqrt{2}\sigma_{\Delta V_n}}\right)\right] \tag{8}$$

where $\Phi_{\Delta V_n}(v)$ is the probability density function of the normally distributed random variable ΔV_n, with variance $\sigma_{\Delta V_n}$.

Finally, this can be rewritten in terms of the equivalent input noise, $\sigma_{V_{in}}$.

$$p = \frac{1}{2}\left[1 + \mathrm{erf}\left(\frac{\Delta V_T}{\sqrt{2}\sigma_{V_{in}}}\right)\right] \tag{9}$$

3.2. ID Cell Reliability

Knowing the relationship between p and ΔV_T allows for the reliability of each ID cell to be determined quickly. The magnitude of ΔV_T necessary for a given reliability threshold ϵ, V_R, can be calculated by inverting Equation (9) with $p = 1 - \epsilon$.

$$V_R = \sqrt{2}\sigma_{V_{in}}\mathrm{erf}^{-1}(1 - 2\epsilon) \tag{10}$$

As described in [17], two tests can then be performed on each ID cell, first by adding a difference voltage V_R between the gates of M1 and M2 during read, such that $V_{G1} - V_{G2} = V_R$. If the result is a "0", then $\Delta V_T < -V_R$. Next, a difference voltage $-V_R$ is applied. If the result is a "1", then $\Delta V_T > V_R$.

The ID cell can then be classified into one of three categories: reliable "1", reliable "0", or unreliable, based on the two tests. The classification system is shown in Table 1.

4. Masked Addressing

Because unreliable ID bits may lead to identification errors, some method for ensuring reliable identification is required. One obvious possibility would be to simply exclude all chips that have unreliable ID bits from use, but this would significantly reduce the yield, particularly in systems employing large numbers of ID bits.

Table 1. Classification of ID cells based on reliability tests, where V_R is the magnitude of the threshold voltage difference required for reliable operation, C is ID code, and M is the ID mask.

$\Delta V_T < -V_R$	$\Delta V_T > V_R$	Classification	C	M
True	False	"0"	0	1
False	False	Unreliable	X	0
False	True	"1"	1	1

An alternative is to record which ID bits of a particular chip are reliable, and then exclude the unreliable ID bits from use during identification. This is accomplished by recording two numbers for every chip: an N-bit ID code C, and an N-bit code mask M. Each bit M_i is equal to "1" if the corresponding ID code bit C_i is reliable, and equal to "0" if C_i is unreliable. The values of C_i and M_i can be determined from the reliability tests as indicated in Table 1.

To identify a particular chip A, an external reader transmits both C_A and M_A. Every chip receives this code and activates its ID generator, which produces a generated code, G, and then tests the following equality

$$C_A \ \& \ M_A = G \ \& \ M_A \tag{11}$$

If this equality is true, the chip can determine it is being addressed. Because the generated code, G_A, will only vary from C_A among the bits excluded by the mask, M_A, this will ensure reliable identification. The protocols for initial chip characterization and subsequent chip identification are given below.

4.1. Chip Characterization

Chip characterization occurs once for each chip, recording C and M, which are necessary to identify the chip in the future.

(1) A single chip A, to be characterized, is placed in range of the external reader. The external reader transmits a characterize signal.

(2) The chip applies a voltage difference ΔV_R to the ID cells, activates the ID generator, and records the output G_+.

(3) The chip applies a voltage difference $-\Delta V_R$ to the ID cells, activates the ID generator, and records the output G_-.

(4) The chip transmits G_+ and G_- to the external reader.

(5) The external reader computes:

$$C_A = \overline{G_+} \ \& \ G_- \tag{12}$$

$$M_A = G_+ \ || \ G_- \tag{13}$$

(6) The external reader stores C_A and M_A for chip A.

4.2. Chip Identification

Chip identification occurs when the system needs to locate a particular chip from a group of chips. Following identification, the chip can reply with a simple acknowledgement, and any other data to be collected.

(1) To identify chip A, the external reader transmits $C_A \ \& \ M_A$ and M_A. All chips within range receive this message.

(2) Each chip activates its ID generator, with zero voltage difference applied to the ID cells, producing a generated code G.

(3) Each chip evaluates the statement

$$C_A \ \& \ M_A = G \ \& \ M_A \tag{14}$$

(4) If the preceding step evaluates to true, the chip concludes it is being addressed.

4.3. Performance Metrics

An effect of this addressing scheme is that the maximum number of chips that can be addressed is reduced from the theoretical maximum of 2^N. If only chips that have a number of unstable bits less than or equal to some maximum value, U, are selected for use, than the maximum number of chips that can be uniquely addressed is

$$S = 2^{N-U} \tag{15}$$

For a chip with U unreliable bits, a false negative can only occur if there is an error among one of the $N - U$ reliable bits. If the probability of an error in one of the reliable bits is ϵ, then the false negative rate is

$$p_{FN} = 1 - (1 - \epsilon)^{N-U} \tag{16}$$

Errors among reliable bits could also cause false positive identifications. Calculating the combined probability of any false positive event is difficult, so we will restrict the calculation to single bit errors, since the probability of E bit errors is ϵ^E, which decreases rapidly for $E > 1$. For $E = 1$, only chips with codes within Hamming distance 1 of the target chip code can cause false positives. In the worst case, for an N bit code with U unreliable bits, there are $(N - U)2^U$ possible chip codes within a Hamming distance of 1. However, this number is usually much larger than the maximum number of uniquely addressable chips, S, given by Equation (15). In actual practice, therefore, there are at most $S - 1$ other chips that could cause false positive errors. The probability of any of these chips causing a false positive is then

$$p_{FP} = 1 - (1 - \epsilon)^{2^{N-U}-1} \tag{17}$$

The total error rate, p_E can then be found

$$p_E = 1 - (1 - p_{FN})(1 - p_{FP}) \tag{18}$$

$$p_E = 1 - (1 - \epsilon)^{(N-U)(2^{N-U}-1)} \tag{19}$$

Restricting the use of chips to those that have a number of unreliable bits less than or equal to U reduces the yield. If the probability that a bit is unreliable is p_U, then the yield is given by

$$Y = \sum_{u=0}^{U} \binom{N}{u} p_U^u (1 - p_U)^{N-u} \tag{20}$$

5. Experimental Results

The proposed circuit was fabricated in a 130 nm CMOS process. A photograph of the manufactured die is shown in Figure 4. As in Figure 1, one column of ID cells shared a single SA, with 32 ID cells per column. A shift register is used to generate the row select signal, and each shift register is shared by 32 columns to for a 32 × 32 array. Fifteen of these arrays are contained on each die, for a total of 15,360 ID bits per die. One 32-bit column, including the SA, occupies an area of 68.5 × 4.3 μm. For measurement, the value of V_{DD} was 400 mV.

Figure 4. Photograph of random ID chip with 15,360 ID bits, fabricated in 130 nm CMOS process.

For each ID cell, the value of p was determined by reading from the cell 1000 times, and counting the number of results equal to "1". To find the value of the threshold voltage mismatch, ΔV_T in the ID cell, the gate of M1 was held fixed, and the gate of M2 was swept in 1 mV increments. At each increment, the ID cell was read 100 times. From these two measurements, the relationship between p and ΔV_T was determined, and plotted in Figure 5. From this data, the value of the input referred voltage noise, $\sigma_{V_{in}}$, was determined to be 1.5 mV.

If a reliability threshold of $\epsilon = 10^{-10}$ is chosen, Equation (10) indicates ID cells with $|\Delta V_T| > 10$ mV will behave reliably. Of the fabricated ID cells examined, 82% satisfy this inequality and can be classified as reliable ($p_U = 0.18$). More generally, Figure 6 shows the relationship between $|\Delta V_T|$ and the reliability threshold, and Figure 7 shows the fraction of unreliable bits for a given reliability threshold.

The location of unreliable cells for an arbitrarily chosen 32×32 array is shown in Figure 8. The unreliable cells are shown in black. The magnitude of ΔV_T is also shown for the same array. This indicates that unreliable bits have random spatial distribution in the array. The measured distribution of ΔV_T is shown in Figure 9, along with a Monte Carlo simulation of the same distribution. This shows the distribution of ΔV_T can be accurately predicted in advance for well modeled processes.

Figure 5. Relationship of ID cell probability, p, to threshold voltage mismatch ΔV_T, measured over 15,360 bits. Equation (9) is also plotted with $\sigma_{V_{in}} = 1.5$ mV.

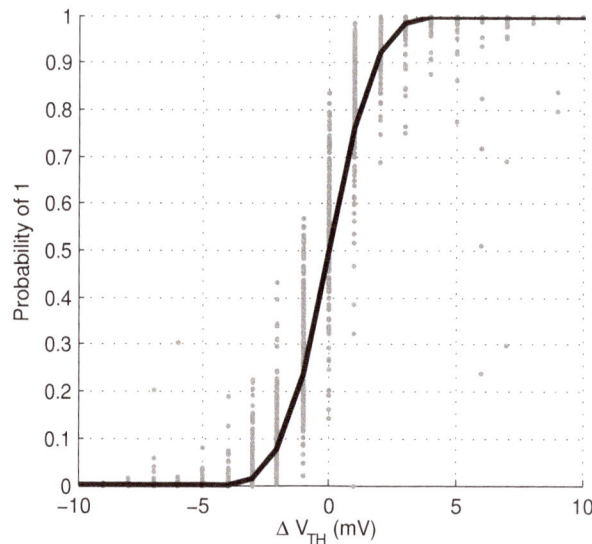

Figure 6. Relationship between ID cell error rate, ϵ, and the magnitude of the threshold voltage mismatch in the ID cell, $|\Delta V_T|$.

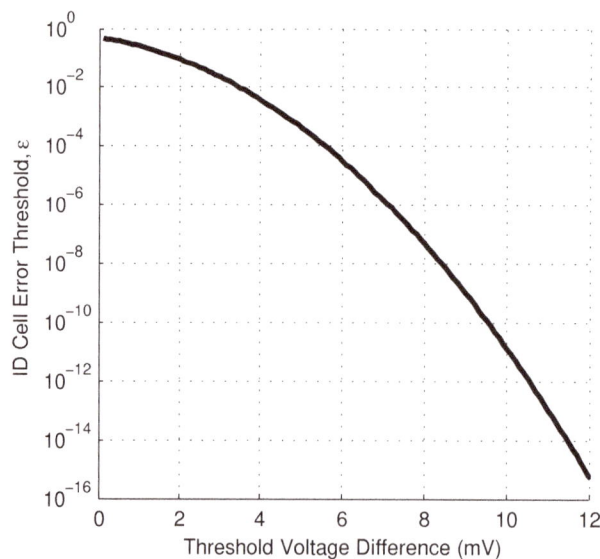

The relationship between $|\Delta V_T|$ and the reliability threshold depends on the value of $\sigma_{V_{in}}$, which will vary in different technologies. Equation (9) shows that the required $|\Delta V_T|$ is linearly related to $\sigma_{V_{in}}$, *i.e.*, a 10$\times$ increase in $\sigma_{V_{in}}$ would require a 10$\times$ increase in the difference threshold $|\Delta V_T|$ to maintain the same error threshold. This may reduce the yield, if the standard deviation of the threshold voltage is not also higher in the alternate technology.

The energy consumption was measured to be 39 fJ/bit at a readout rate of 40 kBps. This is 23$\times$ less than the next lowest published value of 930 fJ/bit [16]. The decreased energy consumption is likely due to the lower supply voltage, sharing of a single SA (which amortizes the SA leakage current), and the use of a shift register for sequential access rather than a row decoder. The effective area per bit (32-bit column and sense amplifier divided by 32), is 9.2 μm^2.

Figure 7. Fraction of manufactured bits that can be expected to be unreliable for a given threshold of reliability, ϵ.

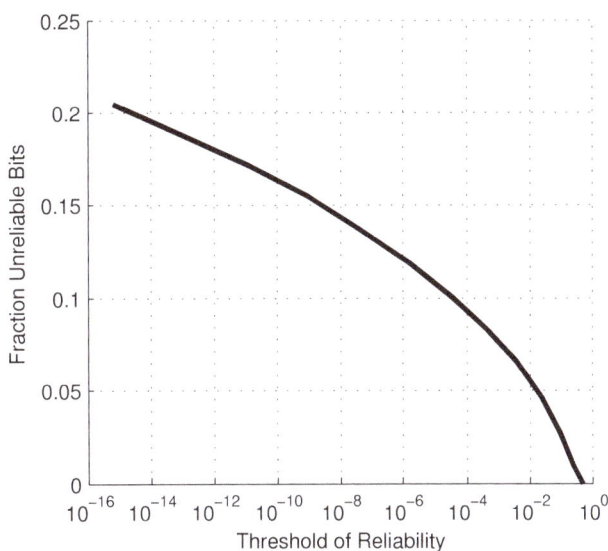

Figure 8. (**A**) Location of unreliable bits ($|\Delta V_T| < 10$ mV), shown in black, in a 32 $\times$ 32 array of random ID cells. (**B**) Magnitude of $|\Delta V_T|$ in the same array, where black is 0 mV and white is 50 mV.

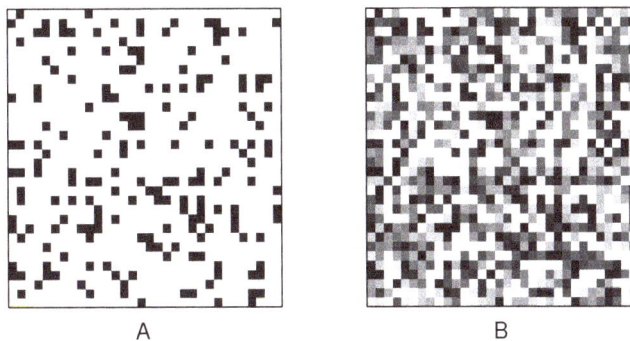

Figure 9. Normalized histogram showing measured and simulated distribution of ΔV_T. Measured sample size is 15,360 bits, and simulated sample size is 10,000 bits.

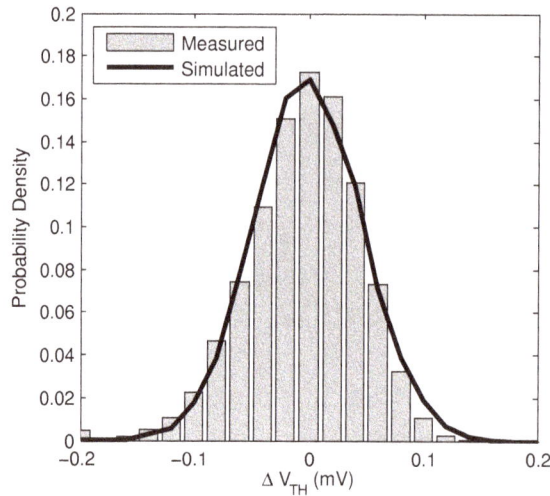

As previously noted, having a non-zero number of unreliable bits, U, reduces the maximum number of addressable chips from the theoretical maximum of 2^N. For this system, with $\sigma_{V_{in}} = 1.5$ mV, a chosen reliability threshold of $\epsilon = 10^{-10}$ and $p_U = 0.18$, and a 99.9% yield, the maximum number of addressable chips *versus* the number of ID bits is shown in Figure 10. This shows reasonable performance as the addressing scheme is scaled up to large systems.

Figure 10. Maximum number of addressable chips under the masked addressing system for a given number of random ID bits, N.

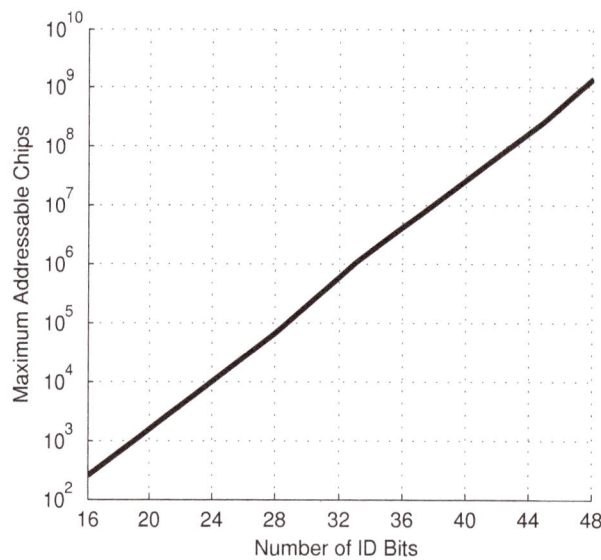

5.1. Temperature Dependence

To evaluate the reliability of the random ID system over temperature, the change in ΔV_T was measured for all the ID cells over 30 °C and 50 °C increases in temperature. We observed that for each ID cell, the temperature dependence of ΔV_T varies, in part proportionally to ΔV_T, and in part randomly between cells. The change in ΔV_T over the observed range is roughly linear with temperature, and so can be expressed as

$$\frac{d\Delta V_T}{dT} = K \tag{21}$$

where K is a random variable, with mean value proportional to ΔV_T:

$$\mu_K = a\Delta V_T \tag{22}$$

The temperature coefficient, a, was found to be $-6.1 \times 10^{-3}\ ^{\circ}\mathrm{C}^{-1}$. Because a is negative, the magnitude of ΔV_T tends to decrease as temperature increases. The standard deviation of K was found to be $\sigma_K = 0.22\ \mathrm{mV}/^{\circ}\mathrm{C}$. A plot of the mean and standard deviation of the temperature dependence is shown in Figure 11, over a range of ΔV_T values.

A possible explanation for the apparent randomness of the temperature dependence is that the devices M1 and M2 in Figure 1 are assumed to differ only in their values of V_T, when in fact other device parameters such as the mobility μ_n, gate oxide capacitance C_{ox}, and effective dimensions W and L will vary as well [19]. For example, in the preceding characterization scheme, if devices M1 and M2 are determined to have equal drain currents when driven with equal gate voltages, it is assumed that their values of V_T are equal. However, it could also be that their values of V_T are unequal, and other devices parameters are unequal in such a way that the drain currents remain equal. These other device parameters will then have their own, unequal temperature dependence [20], resulting in the apparent random temperature dependence. A possible improvement to this work would be a more elaborate characterization scheme in which the mismatch between devices M1 and M2 could be more completely determined, although this would increase the complexity of the characterization scheme.

Figure 11. Mean and standard deviation of the change in ΔV_T per 1 $^{\circ}$C change in temperature, derived from measurement of 15,360 bits of a single chip.

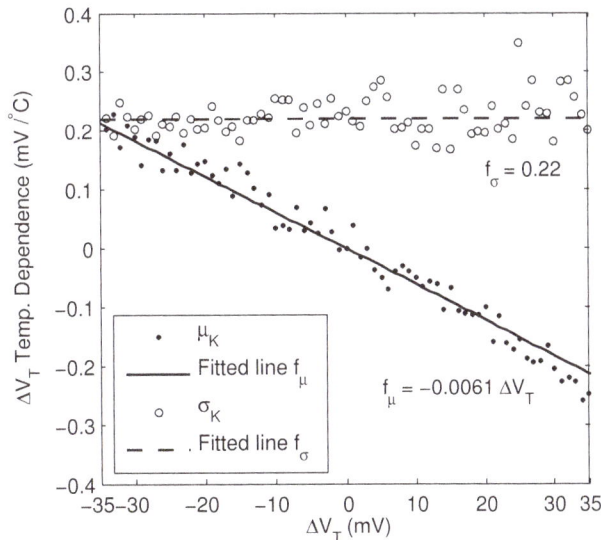

The practical effect of this is that a larger value of V_R must be selected for reliable operation, due to the combined effects of electronic noise and temperature dependence. For positive values of ΔV_T, and if all temperature dependencies are assumed to fall within 6σ, then the worst case temperature shift from ΔV_T to $\Delta V_T'$ over an increase in temperature ΔT is

$$\Delta V_T' - \Delta V_T = (a\Delta V_T - 6\sigma_K)\Delta T \tag{23}$$

Rearranging Equation (23) and substituting $\Delta V_T' = V_R$ and $\Delta V_T = V_R'$ gives the value of V_R' measured at nominal temperature that is required to ensure $|\Delta V_T'| \geq V_R$ after a temperature increase of ΔT.

$$\Delta V_R' = \frac{V_R + 6\sigma\Delta T}{1 + a\Delta T} \tag{24}$$

That is, by only selecting ID cells with $|\Delta V_T| \geq V_R'$, it is ensured that after a ΔT increase in temperature, the new threshold voltage difference will satisfy the original reliability requirement, $|\Delta V_T'| \geq V_R$. For example, using the previously determined value of $V_R = 10$ mV, and assuming a possible temperature increase of 10 °C, Equation (24) indicates that $V_R' = 25$ mV. This increases the fraction of unreliable bits to $p_U = 0.41$.

This method could be employed by determining the parameters a and σ of Equation (24) once for a particular manufacturing process. The characterization of individual chips would then not require any extra measurements beyond the two already required by the characterization scheme, and these measurements do not have to be taken at a particular temperature. This method is, however, inefficient due to the constraint imposed by Equation (23), which assumes the worst case temperature dependence. This is highly unlikely for any given ID cell; a large fraction of cells characterized in this way as unreliable would in fact behave reliably. An obvious alternative would be simply to characterize each cell twice, once at each extreme of the expected operating temperature range, and only select cells that have the same classification at both extremes as reliable. This is more efficient, in the sense that the fraction of unreliable bits will be smaller, but requires a temperature controlled environment for characterizing each chip, which is frequently expensive and time-consuming to employ.

5.2. Addressing Parameter Selection

The selection of the particular parameters of the addressing scheme depends heavily on the particular application. In most cases, the specifications will require the selection of three minimum requirements: (1) address space size (S); (2) yield (Y); and (3) error rate (p_E). After these requirements are specified, the other parameters of the masked addressing scheme can be determined. The parameters of a hypothetical system are determined here using experimentally derived data, to give an indication of the performance of this work.

Consider a system with the specifications $S = 1000$, $Y = 99.9\%$, $p_E = 10^{-6}$, and a possible $\Delta T = 10$ °C. Given the requirement $S = 1000$ and Equation (15), the following constraint is imposed

$$N - U \geq 10 \tag{25}$$

Given this constraint, a lower limit for N can be found using Equation (20), with $Y = 0.999$ and $p_U = 0.41$. The smallest value that satisfies these conditions is $N = 31$, setting $U = 21$. Finally the total error rate can be found using Equation (19) to be $p_E = 1.03 \times 10^{-7}$.

6. Conclusions

This work presents a circuit and addressing scheme that can be used to remotely identify integrated circuits using their inherent manufacturing variation. Unlike other identification techniques based on manufacturing variation, identification can be done on-chip, which reduces the amount of data that must be transmitted by the chip. This is made possible by characterizing the reliability of the random ID bits, and then developing an addressing scheme that is insensitive to errors among the unreliable bits. Our approach uses 39 fJ/bit per bit, which is 23× less energy than prior art, and includes a mathematical treatment of the reliability of such a system. Additionally, this circuit could be used as part of a low-energy secret key generator for cryptographic applications.

Author Contributions

Jonathan F. Bolus was responsible for authoring this paper, the design and test of the circuits described here, the development of the masked addressing system and the reliability analysis. Travis N. Blalock and Benton H. Calhoun helped to guide this research, review the proposed circuits, develop the masked addressing system, and edit this paper.

Conflicts of Interest

The authors declare no conflict of interest.

References

1. Cong, P.; Ko, W.H.; Young, D.J. Wireless batteryless implantable blood pressure monitoring microsystem for small laboratory animals. *IEEE J. Solid-State Circuits* **2010**, *10*, 243–254.

2. Law, M.K.; Bermak, A.; Luong, H.C. A Sub-uW embedded CMOS temperature sensor for RFID food monitoring application. *IEEE J. Solid-State Circuits* **2010**, *45*, 1246–1255.

3. Yakovlev, A.; Pivonka, D.; Meng, T.; Poon, A. A mm-Sized Wirelessly Powered and Remotely Controlled Locomotive Implantable Device. In Proceedings of the 2012 IEEE International Solid-State Circuits Conference Digest of Technical Papers (ISSCC), San Francisco, CA, USA, 19–23 February 2012.

4. Liao, Y.; Yao, H.; Lingley, A.; Parviz, B.; Otis, B.P. A 3-μW CMOS glucose sensor for wireless contact-lens tear glucose monitoring. *IEEE J. Solid-State Circuits* **2012**, *47*, 335–344.

5. Kuhl, M.; Gieschke, P.; Rossbach, D.; Hilzensauer, S.A.; Panchaphongsaphak, T.; Ruther, P.; Lapatki, B.G.; Paul, O.; Manoli, Y. A wireless stress mapping system for orthodontic brackets using CMOS integrated sensors. *IEEE J. Solid-State Circuits* **2013**, *48*, 2191–2202.

6. Lofstrom, K.; Daasch, W.R.; Taylor, D. IC Identification Circuit Using Device Mismatch. In Proceedings of the 2000 IEEE International Solid-State Circuits Conference, San Francisco, CA, USA, 7–9 February 2000.

7. Gassend, B.; Clarke, D.; van Dijk, M.; Devadas, S. Silicon Physical Random Functions. In Proceedings of the 9th ACM Conference on Computer and Communications Security, Washington, DC, USA, 18–22 November 2002.

8. Lee, J.W.; Lim, D.; Gassend, B.; Suh, G.E.; van Dijk, M.; Devadas, S. A Technique to Build a Secret Key in Integrated Circuits for Identification and Authentication Applications. In Proceedings of the 2004 Symposium on VLSI Circuits, Honolulu, HI, USA, 17–19 June 2004.

9. Suh, E.G.; Devadas, S. Physical Unclonable Functions for Device Authentication and Secret Key Generation. In Proceedings of the 44th ACM Annual Design Automation Conference, San Diego, CA, USA, 4–8 June 2007.

10. Holcomb, D.; Burleson, W.; Fu, K. Power-up SRAM state as an identifying fingerprint and source of true random numbers. *IEEE Trans. Comput.* **2009**, *58*, 1198–1210.

11. Fujiwara, H.; Yabuuchi, M.; Nakano, H.; Kawai, H.; Nii, K.; Arimoto, K. A Chip-ID Generating Circuit for Dependable LSI using Random Address Errors on Embedded SRAM and On-Chip Memory BIST. In Proceedings of the 2011 Symposium on VLSI Circuits Digest of Technical Papers, Honolulu, HI, USA, 15–17 June 2011; pp. 76–77.

12. Rosenblatt, S.; Fainstein, D.; Cestero, A.; Safran, J.; Robson, N.; Kirihata, T.; Iyer, S.S. Field tolerant dynamic intrinsic chip ID using 32 nm high-K/metal gate SOI embedded DRAM. *IEEE J. Solid-State Circuits* **2013**, *48*, 940–946.

13. Lim, D.; Lee, J.W.; Gassend, B.; Suh, E.; van Dijk, M.; Devadas, S. Extracting secret keys from integrated circuits. *IEEE Trans. Large Scale Integr. (VLSI) Syst.* **2005**, *13*, 1200–1205.

14. Niewenheuis, B.; Blanton, R.D.; Bhargava, M.; Mai, K. SCAN-PUF: A Low Overhead Physically Unclonable Function from Scan Chain Power-Up States. In Proceedings of the 2013 IEEE International Test Conference, Anaheim, CA, USA, 6–13 September 2013.

15. Hirase, J.; Furukawa, T. Chip Identification Using the Characteristic Dispersion of Transistor. In Proceedings of the 14th Asian Test Symposium, Calcutta, India, 18–21 December 2005.

16. Su, Y.; Holleman, J. A digital 1.6 pJ/bit chip identification circuit using process variations. *IEEE J. Solid-State Circuits* **2008**, *43*, 69–77.

17. Bhargava, M.; Mai, K. An Efficient Reliable PUF-Based Cryptographic Key Generator in 65 nm CMOS. In Proceedings of the 2014 Conference on Design, Automation & Test in Europe, Dresden, Germany, 24–28 March 2014.

18. Dell, B.; Bolus, J.F.; Blalock, T.N. An Automated Unique Tagging System Using CMOS Process Variation. In Proceedings of the 17th ACM Great Lakes Symposium, Stresa-Lago Maggiore, Italy, 11–13 March 2007.

19. Pelgrom, M.J.M.; Duinmaijer, A.C.J.; Welbers, A.P.G. Matching properties of MOS transistors. *IEEE J. Solid-State Circuits* **1989**, *24*, 1433–1439.

20. Vadasz, L.; Grove, A.S. Temperature dependence of MOS transistor characteristics below saturation. *IEEE Trans. Electron Devices* **1966**, *13*, 863–866.

10

Untrimmed Low-Power Thermal Sensor for SoC in 22 nm Digital Fabrication Technology

Ro'ee Eitan and Ariel Cohen *

Intel Israel LTD., S.B.I. Park Har Hotzvim, CFF8, Jerusalem 91031, Israel;
E-Mail: roee.eitan@intel.com

* Author to whom correspondence should be addressed; E-Mail: ariel.cohen@intel.com

External Editor: Alexander Fish

Abstract: Thermal sensors (TS) are essential for achieving optimized performance and reliability in the era of nanoscale microprocessor and system on chip (SoC). Compiling with the low-power and small die area of the mobile computing, the presented TS supports a wide range of sampling frequencies with an optimized power envelope. The TS supports up to 45 K samples/s, low average power consumption, as low as 20 μW, and small core Si area of 0.013 mm^2. Advanced circuit techniques are used in order to overcome process variability, ensuring inaccuracy lower than ±2 °C without any calibration. All this makes the presented thermal sensor a cost-effective, low-power solution for 22 nm nanoscale digital process technology.

Keywords: thermal sensor; sigma-delta Analog to Digital Converter (ADC); low-power; untrimmed temperature sensor; nanoscale digital process

1. Introduction

Integrated thermal sensor (TS) circuits have become key elements in high performance systems, especially in processors and system on chip (SoC). Such applications require a relatively high precision thermal sensor (typically ±3 °C in a wide temperature range), in order to achieve high reliability and performance [1,2]. The conventional method of thermal sensing is based on diode connected Bipolar Junction Transistor (BJT) voltage temperature dependence [2–7]. A base-emitter voltage equation of BJT transistor is (to first order approximation):

$$V_{BE} = \frac{nkT}{q} Ln(\frac{I}{I_0})$$ (1)

where n is the ideality factor of the diode, k is Boltzmann's constant, T is the absolute temperature, q is the electron charge, I_0 is the diode's saturation current and I is the current through the diode. This voltage is inversely proportional to the temperature, because of the strong temperature dependence of the saturation current I_0. The diode voltage is marked as V_{BE}, since a diode connected BJT transistor is used in order to implement a diode, from this point onwards when referring to "diode" the meaning is diode connected BJT.

The differential voltage between the two diodes is:

$$\Delta V_{BE} = V_{BE2} - V_{BE1} = \frac{nkT}{q}[\ln(\frac{I_2}{I_0}) - \ln(\frac{I_1}{I_0})] = \frac{nkT}{q}\ln(M)$$ (2)

where M is the ratio between the diodes' current density. This voltage is directly proportional to the temperature.

Several methods have previously been used to read out the diodes' voltage and convert it into a temperature reading. One of the popular implementations, used at industry, finds the crossing point between the amplified differential voltage of the two diodes, which is directly proportional to the temperature, and the voltage of a single diode, which is inversely proportional to the temperature (see Figure 1) [8]. For a given temperature, there will be exactly one crossing point, which exists only for a specific diodes' current. This current is controlled by a Digital to Analog Converter (DAC). A controller sweeps across the current DAC until this point is found by a comparator. The DAC code at this point represents the evaluated temperature [9]. This architecture is relatively limited in linearity and accuracy over a wide temperature range.

Moreover, there is a variance in the slope of the linear functions, due to sensitivity to process variations. The slope initial variation is large, requiring two-temperature calibration in order to achieve the required accuracy. Performing two temperature calibrations during High Volume Manufacturing (HVM) tests is considered costly.

Several motivations drove and guided the design of the temperature sensor presented in this paper: (a) to provide the best possible accuracy, as accurate temperature sensing extends performance of SoCs; (b) to utilize advanced circuit techniques to overcome process variance effects and, thus, to get a robust and accurate TS; (c) to reduce the products' HVM calibration process cycle to a single-temperature calibration or to no calibration; (d) to achieve goals (a) and (b) with the smallest possible power and area penalty; and (e) to be configurable in order to meet a wide range of power *versus* sampling rate optimized envelop for different requirements of several projects.

Figure 1. Operation principle of thermal sensor (CD—current density; a—temperature coefficient of ΔV_{BE}).

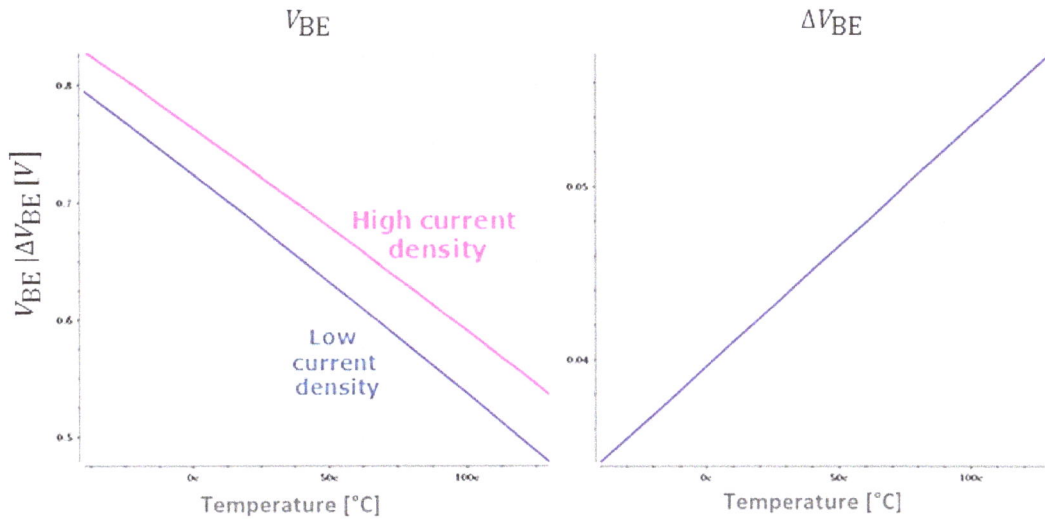

$$\Delta V_{BE} = (\frac{kT}{q})\ln(\frac{highCD}{lowCD}) \quad Sensitivity(a) = 138\frac{\mu V}{^\circ C}$$

$$\Delta V_{BE} = aT \qquad \textit{for} \text{ current density ratio of 5}$$

2. 22 nm Sigma-Delta Based Thermal Sensor Architecture

Better linearity is achieved using the Analog to Digital Converter (ADC) based thermal sensor [3,4]. In this architecture, the differential voltage is amplified and converted into a digital word by an ADC. A bandgap voltage reference circuit (BG) can be used in order to supply a precision reference voltage for the ADC. This operating mode is called Vref mode. Another operating mode is creating the reference voltage internally in the TS. This is referred as Ratiometric mode (Figure 2). The ADC-based architecture has very good linearity over a wide range of temperatures. The accuracy of this architecture can be improved by using a calibration process during production. This architecture may produce accuracy of better than ±0.5 °C with old processes [3,4,6,7].

The thermal sensor presented here was fabricated in the 22 nm process. The architecture includes an advanced sense stage combined with circuits that deal with process variations (chopping and dynamic element matching). A second order sigma-delta (SD) ADC is used both as an amplifier and as a converter that translates the temperature readings to a digital word. The SD converter includes correlated double sampling which improves its 1/f attenuation and offset performance and a custom made digital block. A synthesized block implements the sigma-delta decimation filters and provides a digital readout of the relative operating temperature as well as temperature alerts. This sigma-delta ADC based TS can be operated in two modes: Reference-based (Vref) and Ratiometric-based (Ratiometric). In the Ratiometric approach, the converter extracts bandgap voltages directly from the sense stage. Stabilization is achieved by feed-forward methodology. The sigma-delta modes are selectable through digital control manipulation only, without additional analog components. A thermal diode was also included within the TS for junction temperature sensing by a ±1 °C sensor, as part of the calibration process.

The TS has two main power modes: high sampling rate and power saving. In the power saving mode, the TS is switched on for a short time that enables reading of accurate temperature and most of the time the TS is off. In this method, the TS probes the temperature in frequencies of 10 to 100 samples/s . In the high sampling rate mode the TS is continuously working with the sigma-delta modulator clocked at a lower frequency and samples the temperature in frequencies of 5 K to 45 K samples/s.

Figure 2. High level diagram of the Thermal Sensor.

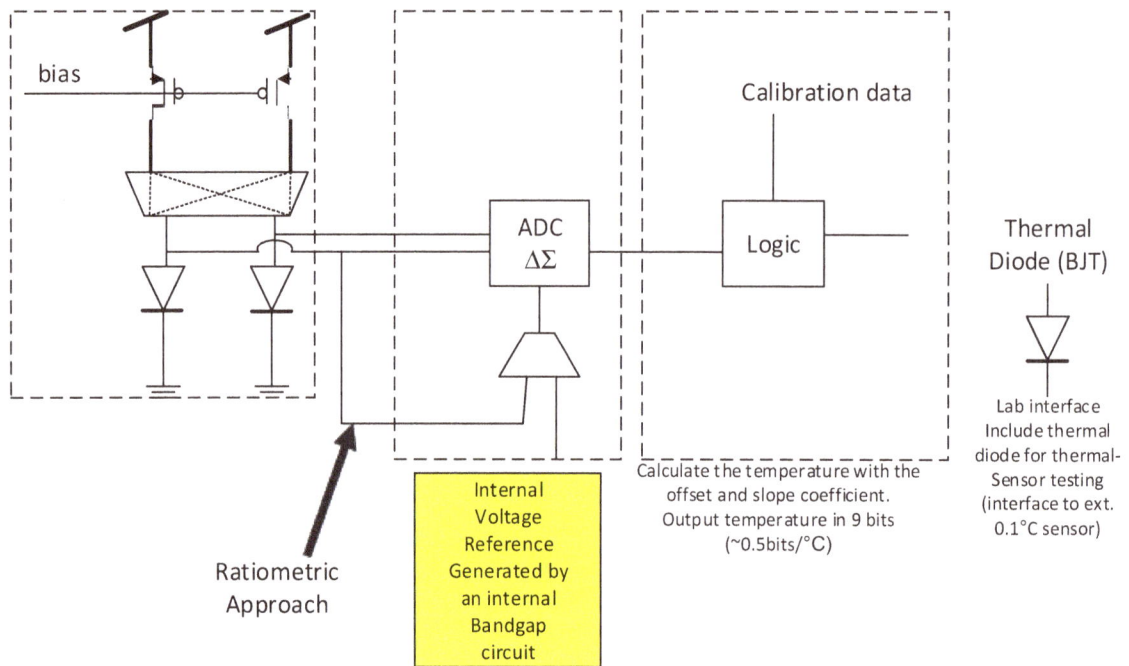

Device mismatches, such as between current mirrors and between diodes, are effectively treated by the advanced dynamic compensation circuit techniques. However, there are a few errors which require compensation during calibration if an improved accuracy is required. The sources of these errors are process variations of diode parameters, such as the ideality factor, variation of the conversion factor of the ADC dominated by capacitor mismatch, and Bandgap reference voltage variation in Vref mode only. The High Volume Manufacture (HVM) calibration result is stored in on-die fuses for use by the thermal sensor.

3. The Sense Stage

The sense stage diagram, presented in Figure 3, converts the junction temperature to a Proportional to Absolute Temperature (PTAT) voltage. It includes a matched pair of diodes and current mirrors with controlled current ratios (of 3.5, 4, 5, 6, 7, and 8). The absolute value of the mirrored current is controlled by an internal resistor and is equal to a reference voltage divided by the resistor value. The absolute current is designed to be a controlled value since it affects the current density within the diodes, thus affecting their performance as thermal sensing elements. In this work, the bias current of a single current source is nominally 15 μA. The current mirrors use gain boosting (regulated cascode) in order to increase current mirror output resistance, Rout, thus, further improves the current accuracy. These circuits paved the way to achieve accurate TS without the need of temperature calibration.

Figure 3. Sense stage block diagram.

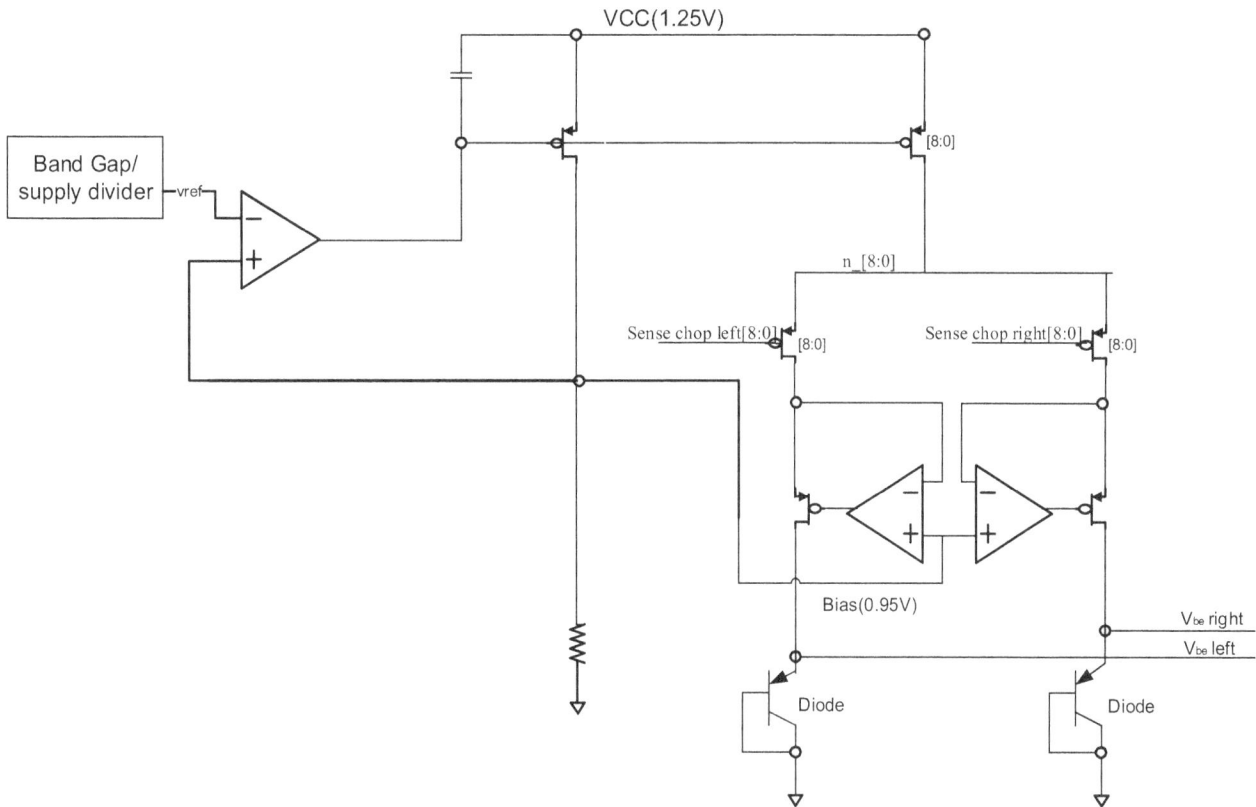

Current ratios are realized by utilizing nine identical current sources and by switching different subsets of the current to either the right or the left diode. In addition, current sources may be switched to a third diode, the waste diode (not shown in the figure), if they are not in use. The waste diode is implemented by driving current into dummies in the matched diode structure, thus avoiding the area penalty of extra circuitry and enable operating of the sense stage in higher frequency.

In order to obtain accurate current ratios resilient to mismatch, Dynamic Element Matching (DEM) is used. The DEM mechanism is realized by cyclically rotating the nine current source connections to the sense diodes and time-averaging the current value of the current mirrors.

Chopping is also used by the sense circuit in order to overcome diode mismatches. After selecting a current ratio, for example 7:2, it may be connected to the diodes in the reverse order as well (e.g., 2:7), thus changing the sign of ΔV_{BE}. This effectively modulates the signal on a carrier frequency, later to be demodulated by the sigma-delta converter. The chopping frequency is identical to the sampling frequency of the sigma-delta ADC so demodulation is back to DC.

From the circuit perspective, the sense stage is implemented with digital transistors on a relatively low power rail (1.25 V). The low threshold voltage of the digital transistors improves the circuits' headroom and improves the matching and Rout of the current mirrors and aids in achieving a good PSRR and an accurate TS in a small die area.

4. The Sigma-Delta Modulator

The sigma-delta modulator is a second order, one bit, switched-capacitor based design. Switched-capacitor circuits are frequently used for sigma-delta designs because they are fully compatible with digital CMOS processes. For low bandwidth applications, the over-sampling ratio (OSR) may be high enough so as not to limit the resolution of the ADC. A second order modulator is a preferred choice because it greatly reduces the stability problems of higher order modulators and decreases idle tone generation. It also lowers requirements for the OSR and thus for the gain of the integrator amplifiers. Integrator parameters (gains) are defined by the ratio of capacitors which is more accurate than absolute values of RC components in continuous time modulators. To reduce power supply noise and other distortions due to the common mode disturbance, a fully differential topology is chosen.

Correlated double-sampling (CDS) techniques and chopping are further used to decrease offset, 1/f noise and inaccuracy effects within the SD ADC.

The principles of the sigma-delta operation in the Vref-based and Ratiometric modes are presented in Figure 4. Charge balance is a fundamental property of any sigma-delta converter. This principle can be used to calculate the average signal transfer function of the modulator. The second stage of the SD-ADC affects mostly the quantization errors and thus does not change the essential transfer function of the ADC.

Figure 4. Sigma-delta Analog to Digital Converter (ADC) operation principle—(**A**) Vref mode; and (**B**) ratiometric mode.

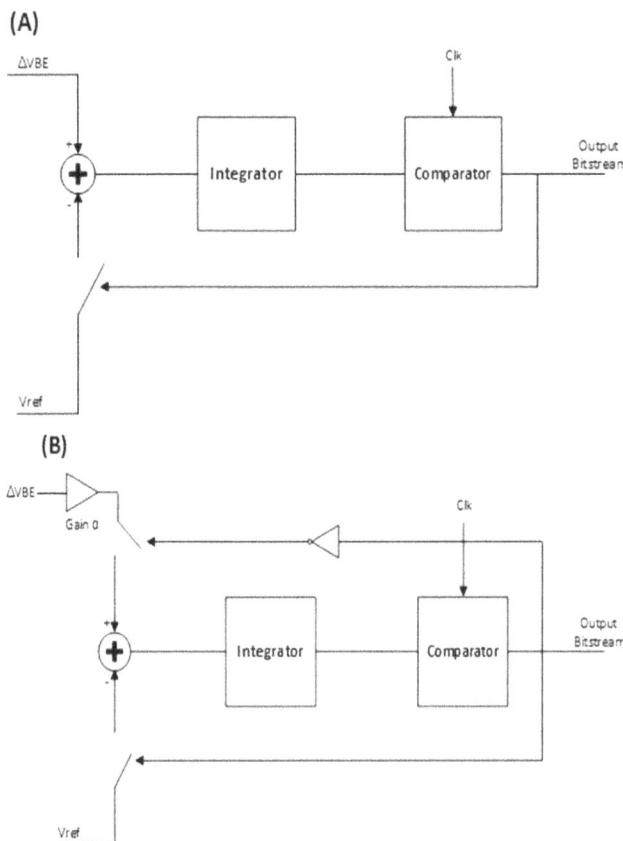

(A)

Charge balance principle:

$$-T_{\text{off}} \times \Delta V_{\text{BE}} = T_{\text{on}} \times (\Delta V_{\text{BE}} - V_{\text{REF}})$$

$$\Rightarrow$$

$$\frac{T_{\text{off}}}{T_{\text{on}}} = \frac{V_{\text{REF}}}{\Delta V_{\text{BE}}} - 1 \Rightarrow$$

$$\left(1 + \frac{T_{\text{off}}}{T_{\text{on}}}\right)^{-1} = \frac{T_{\text{on}}}{T_{\text{on}} + T_{\text{off}}} = \frac{T_{\text{on}}}{T} = \frac{\Delta V_{\text{BE}}}{V_{\text{REF}}}$$

(B)

Charge balance principle:

$$T_{\text{off}} \times \alpha \times \Delta V_{\text{BE}} = -T_{\text{on}} \times (-V_{\text{BE}})$$

$$\Rightarrow$$

$$\frac{T_{\text{off}}}{T_{\text{on}}} = \frac{V_{BE}}{\alpha \Delta V_{\text{BE}}} \Rightarrow$$

$$\left(1 + \frac{T_{\text{off}}}{T_{\text{on}}}\right)^{-1} = \frac{T_{\text{on}}}{T_{\text{on}} + T_{\text{off}}} = \frac{\alpha \Delta V_{\text{BE}}}{V_{\text{BE}} + \alpha \Delta V_{\text{BE}}}$$

Charge-balance mathematics are applied for demonstration to the reference-based sigma-delta, showing how the voltage ratio ($\Delta V_{\text{BE}}/V_{\text{REF}}$) is converted into an average duty ratio (or "1" density ratio) by the converter.

Applying the same principle to the ratiometric construction shows how the "1" density ratio in the bitstream of the Ratiometric modulator is caused by the ratio:

$$\frac{\alpha \Delta V_{BE}}{V_{BE} + \alpha \Delta V_{BE}} \tag{3}$$

where α is the preamp gain and ΔV_{BE} and V_{BE} are the diodes' voltages generated by the sense stage. Proper selection of α will cause the denominator to balance the complementary to absolute temperature (CTAT) behavior of V_{BE} against the proportional to absolute temperature (PTAT) behavior of ΔV_{BE}, achieving the reference Vref within the sigma-delta converter itself. This phenomena is named Ratiometric sigma-delta [6,7], and is used in the thermal sensor in order to omit the bandgap from the sense path (utilizing instead the natural "bandgap" that is already embedded in the sense stage).

The architecture of the sigma-delta converter is presented in Figure 5. The architecture is based on the converter presented by Pertijs and al [7]. Each amplifier constitutes an integrator, thus this circuit is a second order design.

The first integrator incorporates an offset cancelling scheme. It stores the offset on the input capacitors with reverse polarity and during the integration phase the stored offset is subtracted from the integrated signal. The input capacitors to the first amplifier are used to implement a controlled gain. If FS3 is off, then the bucket capacitor of the integrator consists of Cs1 only. Therefore the gain of the integrator would be Cs1/Cint1. However, if Cs2 is connected, the gain would be (Cs1 + Cs2)/Cint1. We use this feature in Vref-mode to boost the ΔV_{BE} signal by a factor of eight, and in the Ratiometric mode as part of the gain α. This eliminates the need for a preamplifier between the sense stage and the ADC converter, thus saving power and area. Additional gain may be obtained, if required, by performing several cycles of integration before the comparator clock (Φ_{cmp}) is asserted.

Figure 5. Second order sigma-delta converter.

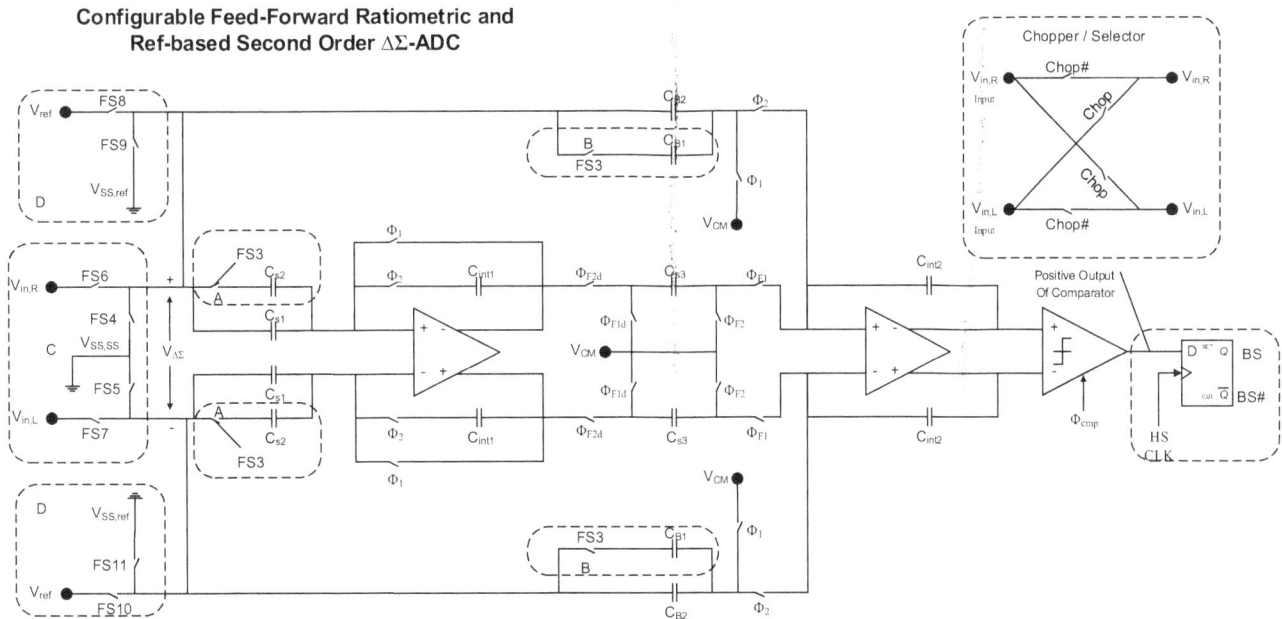

The second stage is a switch capacitor integrator. Feed-forward path stabilization causes low swing at the opamps' outputs so less distortion is expected. No DC component above CM voltage exists at the first integrator output. However, the feed-forward method makes the transfer function depend on capacitors' mismatch.

The ADC is controlled by several clocks ($\Phi1$, $\Phi1t$ and $\Phi2$, $\Phi2t$ and their delayed versions) and functional switch controls: FS3-FS11. This set of control signals is generated by a custom logic circuit within the ADC, Which selects values for the functional switches in two different modes of operation: Vref-based and Ratiometric-based converters.

The bit stream output feeds a three-stage (CIC) filter that removes high-frequency noise and frames the sampled signal into nine bit samples.

5. The TS Top Hierarchy and Layout

As mentioned above, this TS IP is designed for SoC projects and therefore it is designed as a self-contained collateral that includes all the infrastructure for supporting the requirements of multiple projects. In addition to the TS analog core circuit of 0.013 mm^2, there are other pieces of collateral, including: (1) a thermal diode used in the calibration process; (2) a logic core required for digital DSP, filtering and processing the SD-ADC output; (3) two analog ports for Si testing that are connected directly to critical points in the circuit or through analog to frequency converters (A2F); (4) clamps and decoupling capacitors; and (5) Jtag interface for independent accessibility. The thermal sensor layout picture and specifications are presented in Figure 6.

Figure 6. The self-contained thermal sensor layout picture including above the TS core area also thermal diode, clamp, supply decoupling and logic.

The analog core works on a 1.25 V although the digital transistors do not support such direct voltage stress. In order to overcome the over stress limitations of the transistors, advanced techniques were used, such as adding protecting devices and start-up circuits. The use of digital transistor with higher supply voltages enables high bandwidth, relatively high sampling rates, enough headroom range and very small die area. The ADC improved performance together with a configurable and fast digital filter, reduces the latency time of temperature readout, and enables the TS to work with a wide envelope of sample frequencies with optimized power options.

6. Results

The suggested SD-ADC based thermal sensor was implemented in a 22 nm process and was verified in two generations of test chips on typical and skew material. The evaluation boards included an adapter for a thermal head, which forces a specific temperature and an external component for checking the internal thermal diode for measuring the junctional temperature. Extracting the temperature inaccuracy of each chip was done in Ratiometric mode at a temperature range of 0–110 °C.

Figure 7A depicts the temperature reading errors of typical dies after one temperature calibration at 100 °C in Ratiometric mode with supply voltage of 1.25 V. The error for a wide range of temperature is smaller than ±1 °C. The results of the same chips only untrimmed (non-calibrated) demonstrates an inaccuracy of smaller than ±1 °C (Figure 7B). Note that these results include the error induced by the measuring procedure done by the thermal diode.

Figure 7. Temperature error *versus* junction temperature of different typical chips: (**A**) calibrated in one temperature and; (**B**) untrimmed.

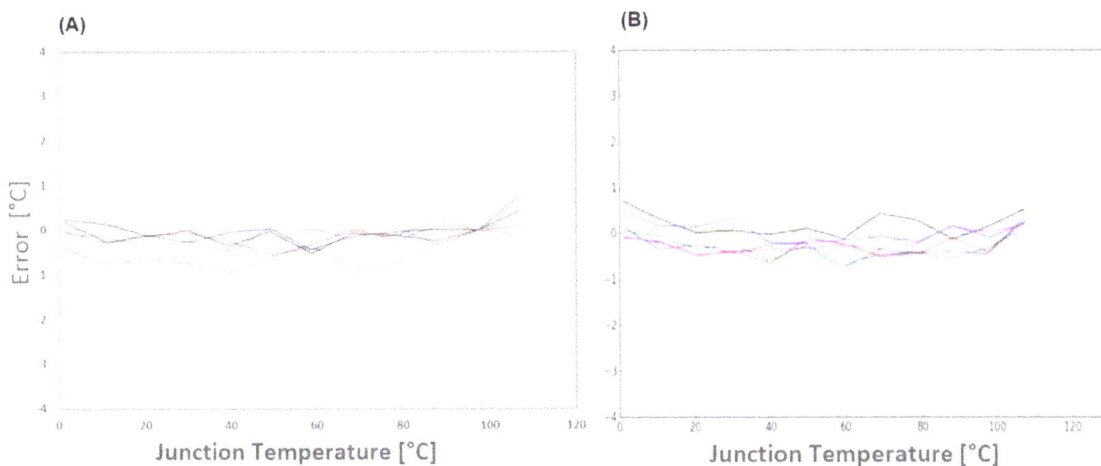

The TS circuit is differential and is very insensitive to power supply variation. The temperature error curves with various supply voltages of 1.25 V ± 5%, and 1.35 V ± 5% are presented in Figure 8 with an inaccuracy of smaller than ±1.8 °C demonstrating a good immunity to supply voltage. In addition, simulations support these results and demonstrate a high PSRR of >50 dB for the ΔV_{BE} signal over a wide frequency range (not presented). The conclusion is that a voltage regulator is not required for this TS supply unlike other TS in the industry that require a voltage regulator [5].

Figure 8. Temperature error *versus* junction temperature of 16 chips in Ratiometric mode with various supply voltages of (**A**) 1.25 V ± 5% and (**B**) 1.35 V ± 5%.

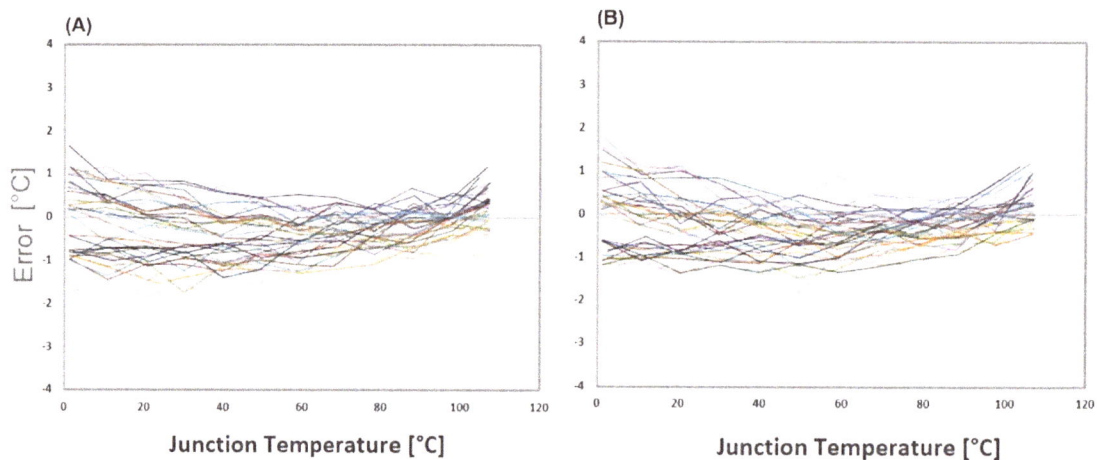

The TS power consumption from the analog supply voltage (1.25 V) appears in Table 1. In the power saving mode in which the sampling rate is low (10–100 Sps), the power consumption is significantly reduced from ~1 mW in high sampling rate mode down to 18 µW. In order to support frequently disabling/enabling the SD ADC, we increased the clock frequency of the sigma-delta modulator. We also reduced the delay on the digital filters by using only three stages of (CIC) filter. A quite short latency for the SD ADC and digital filter was achieved of approximately 40 µs, while maintaining the ADC SNR. The power numbers for both high sampling rate mode (done for the max 45 KSps) and power saving mode are calculated over 16 different chips. The "typical" results represent the average results obtained from typical die, while the "fast" results represent the absolute highest consumption measured for "fast" skew SI.

Table 1. Thermal Sensor power consumption for different sampling rates. "Typical" are the average numbers for typical die where "fast" results represent the absolute highest consumption measured for "fast" skew SI.

Power mode	Corner/Skew	Temp (C°)	Analog supply (V)	Analog current consumption (µA)	Analog power consumption (µW)
Power saving mode (10 Sps)	Typical	60	1.25	14	18
Power saving mode (10 Sps)	Fast	110	1.31	15	20
Power saving mode (100 Sps)	Typical	60	1.25	52	65
Power saving mode (100 Sps)	Fast	110	1.31	60	79
High sampling rate	Typical	60	1.25	780	975
High sampling rate	Fast	110	1.31	920	1205

Finally, the untrimmed inaccuracy, without using any calibration, is quite low (Figure 9) paving the way for using this TS without calibration for most SoC products. The error for 29 devices is ±2 °C in Ratiometric mode (Figure 9A). The average temperature error is <0.5 °C and the resolution is

±0.25 °C without any averaging of reading samples. The one sigma error is less than 1 °C (Figure 9B) and the three sigma error is less than ± 3 °C. Such inaccuracy is close to the results achieved by a practical calibration process such as using thermal diode. It means that calibration is unnecessary.

Figure 9. Untrimmed inaccuracy of thermal sensors in Ratiometric mode. (**A**) Temperature error *versus* junction temperature without calibration for 29 devices; (**B**) Statistics of errors presented in A. The yellow rectangle indicates the average error in a specific temperature and the error bar indicates the standard deviation (1σ) of the error.

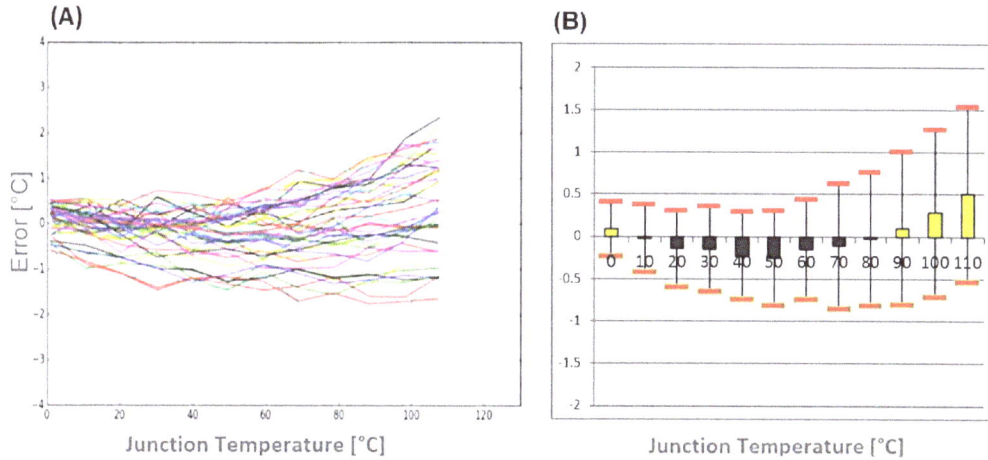

In Table 2 you can see a comparison made with other TS available in the literature. The TS described in this article shows a very good performance in power, area and speed while maintaining good accuracy even without calibration.

Table 2. Comparison with other TSs in the literature. Abs—absolute temperature error.

Reference	Sebastiano [3]	Souri [4]	Lakdawala [5]	Shor [2]	"High Sampling Rate"	"Power Saving Mode"
Source/Year	JSSC 2010	JSSC 2013	JSSC 2009	JSSC 2013	This work	This work
Technology	65 nm	0.16 μm	32 nm	22 nm	22 nm	22 nm
Area (mm²)	0.1	0.08	0.02	0.0061	0.013	0.013
Voltage Regulator	not required	not required	required	not required	not required	not required
Calibration	not required	not required	not required	required	not required	not required
Supply Voltage (V)	1.2–1.3	1.5–2	1.05	1.35	1.25	1.25
Current	8.3 μA	3.4 μA	1.5 mA	1.03 mA	780 μA	15–60 μA
Speed (samples/s)	2.2	10–188	1 K	2 K–20 K	5 K–45 K	10–100
Energy/Sample (J)	4.5 μ	27 n–440 n	1.6 μ	69.5 n–0.69 μ	30 n–270 n	750 n–1.88 μ
Accuracy with/out calibration (°C)	0.2/0.5 (3Σ)	0.15[0.25]/ 0.6[0.8] (3Σ)	/5 (abs)	1.5 (abs)	1.8/2 (abs)	1.8/2 (abs)
Range	−70 to 125	−55 to 125	−10 to 110	−10 to 100	−10 to 110	−10 to 110
Accuracy FOM (J%²) *	47.3 n	0.75 n	27.7 μ	679 n	67.5 n	1.6 μ

* Where it was possible the values were taken from the article. The rest calculated by using the equation at [4].

7. Summary

In this paper we presented a precise low power integrated thermal sensor. The architecture uses advanced features such as dynamic element matching, precise regulated cascode current mirror, chopping, and correlated double sampling to overcome accuracy and matching issues. Silicon validation performed demonstrates immunity to process variation and an error $<\pm1.8$ °C for a range of 0–110 °C with a single temperature calibration.

Some of the configurability options were demonstrated, showing reference-based and Ratiometric operation. In the Ratiometric mode, a reference may not be needed as it is extracted by the ADC from the sense stage itself. Many additional configuration options, such as remote diode interfaces, signal path gain control, diode current ratio control and digital post processing options make this architecture a fertile testing ground for the exploration of temperature sensing techniques.

The untrimmed inaccuracy of the thermal sensor is less than ±3 °C (3σ) across the entire range. Reducing the calibration process will significantly reduce the tester cost for SoC products.

Over all, the silicon results show a unique ability to trade between high sampling rates and low power consumption up to 18 µW while retaining high accuracy without calibration in a small die area.

Acknowledgments

The authors would like to acknowledge the following team members from Intel: Vitali Rahinski, Yosi Sanhedrai, Elad Kuperberg, Tomer Fael, Guillermo Libashevsky, Gilad Peleg and Michael L Tollen who contributed a lot for the design and layout of the thermal sensor and also to Emanuel Natanov and Guy Avigad for leading the post-Si validation.

Author Contributions

Roee Eitan was responsible for the design and testing of the circuits described here. Ariel Cohen guided this research, reviewed the proposed circuits and edited this paper.

Conflicts of Interest

The authors declare no conflict of interest.

References

1. JEDEC Standard. *FBDIMM: Advanced Memory Buffer (AMB)*; JESD82–20; JEDEC Solid State Technology Association: Arlington, VA, USA, 2007.
2. Shor, J.S.; Luria, K. Miniaturized BJT-Based Thermal Sensor for Microprocessors in 32- and 22-nm Technologies. *IEEE J. Solid-State Circuits* **2013**, *48*, 2860–2867.
3. Sebastiano, F.; Breems, L.; Makinwa, K.A.A.; Drago, S.; Leenaerts, D.M.W.; Nauta, B. A 1.2-V 10-µW NPN-based temperature sensor in 65-nm CMOS with an inaccuracy of 0.2 °C (3σ) from 70 °C to 125 °C. *IEEE J. Solid-State Circuits* **2010**, *45*, 312–313.

4. Souri, K.; Chae, Y.; Makinwa, K.A.A. A CMOS temperature sensor with a voltage-calibrated inaccuracy of ±0.15 °C (3σ) from 55 °C to 125 °C. *IEEE J. Solid-State Circuits* **2013**, *48*, 292–301.

5. Lakdawala, H.; Li, Y.W.; Raychowdhury, A.; Taylor, G.; Soumyanath, K. A 1.05 V 1.6 mW 0.45 °C 3σ-resolution ΔΣ-based temperature sensor with parasitic-resistance compensation in 32 nm CMOS. *IEEE J. Solid-State Circuits* **2009**, *44*, 340–341.

6. Pertijs, M.A.; Makinwa, K.A.A.; Huijsing, J. A CMOS smart temperature sensor with a 3 sigma inaccuracy of 0.1 C from 55 C to 125 C. *IEEE J. Solid-State Circuits* **2005**, *40*, 2805–2815.

7. Pertijs, M.A.; Niederkorn, A.; Ma, X.; McKillop, B.; Bakker, A.; Huijsing, J.H. A CMOS smart temperature sensor with a 3 sigma inaccuracy of 0.5 °C from 50 °C to 120 °C. *IEEE J. Solid-State Circuits* **2005**, *40*, 454–461.

8. Wang, G.; Meijer, G.C.M. The temperature characteristics of bipolar transistors fabricated in CMOS technology. *Sens. Actuators A* **2000**, *87*, 81–89.

9. Duarte, D.E.; Geannopoulos, G.; Mughal, U.; Wong, K.L.; Taylor, G. Temperature Sensor Design in a High Volume Manufacturing 65 nm CMOS Digital Process. In Proceedings of the IEEE 2007 Custom Integrated Circuits Conference, San Jose, CA, USA, 16–19 September 2007.

11

Extensionless UTBB FDSOI Devices in Enhanced Dynamic Threshold Mode under Low Power Point of View [†]

Katia Regina Akemi Sasaki [1,*], Marc Aoulaiche [2], Eddy Simoen [2], Cor Claeys [2,3] and Joao Antonio Martino [1]

[1] LSI/PSI—Integrated Systems Laboratory/Department of Electronic Systems Engineering, University of Sao Paulo, Av. Prof. Luciano Gualberto, trav.3, n.158, Sao Paulo 05508-010, Brazil; E-Mail: martino@lsi.usp.br

[2] IMEC—Interuniversity Microelectronic Centre, Kapeldreef 75, B-3001 Leuven, Belgium; E-Mails: Marc.Aoulaiche.ext@imec.be (M.A.); Eddy.Simoen@imec.be (E.S.); Cor.Claeys@imec.be (C.C.)

[3] Electrical Engineering Department, KU Leuven, Kasteelpark Arenberg 10, B-3001 Leuven, Belgium

[†] The original of this paper had been presented in IEEE S3S Conference 2014.

[*] Author to whom correspondence should be addressed; E-Mail: katia@lsi.usp.br

Academic Editors: David Bol and Steven A. Vitale

Abstract: This work presents an analysis about the influence of the gate and source/drain underlap length (L_{UL}) on UTBB FDSOI (UltraThin-Body-and-Buried-oxide Fully-Depleted-Silicon-On-Insulator) devices operating in conventional ($V_B = 0$ V), dynamic threshold (DT, $V_B = V_G$), and the enhanced DT (eDT, $V_B = kV_G$) configurations, focusing on low power applications. It is shown that the underlap devices present a lower off-state current (I_{OFF} at $V_G = 0$ V), lower subthreshold swing (S), lower gate-induced drain leakage (GIDL), higher transconductance over drain current (gm/I_D) ratio and higher intrinsic voltage gain ($|A_V|$) due to their longer effective channel length in weak inversion and lower lateral electric field, while the eDT mode presents higher on-state current (I_{ON}) with the same I_{OFF}, lower S, higher maximum transconductance (gm_{max}), lower threshold voltage (V_T), higher gm/I_D ratio and higher $|A_V|$ due to the dynamically reduced threshold voltage and stronger transversal electric field.

Keywords: dynamic threshold; UTBB; underlap; low power applications

1. Introduction

Thanks mainly to its better back gate coupling, ultrathin-body-and-buried-oxide fully-depleted-silicon-on-insulator (UTBB FDSOI) devices present better performance for sub 28 nm technology nodes such as lower short channel effects (SCE) and better threshold voltage control, while keeping the planar structure [1–5]. Also, they can be used at high frequencies with a lower back gate leakage, lower voltage operation and better power efficiency [4,6]. Moreover, by using a Ground Plane (GP) implantation, the threshold voltage can be adjusted without increasing the channel doping, which avoids mobility degradation and random-dopant fluctuations [7]. This region diminishes the depletion thickness underneath the buried oxide layer, reducing the source and drain electric field that reaches the channel through the buried oxide [7–9].

In circuits, a reduced minimum energy consumption has been demonstrated, which occurs at a lower supply voltage [6], when compared to bulk CMOS devices. This minimum energy consumption is a tradeoff between the leakage and the active current, as indicated in Equation (1) and Figure 1 [10].

$$P = \overbrace{CV_{DD}^{2}f}^{Active} + \overbrace{I_{leak}V_{DD}}^{Leakage} \tag{1}$$

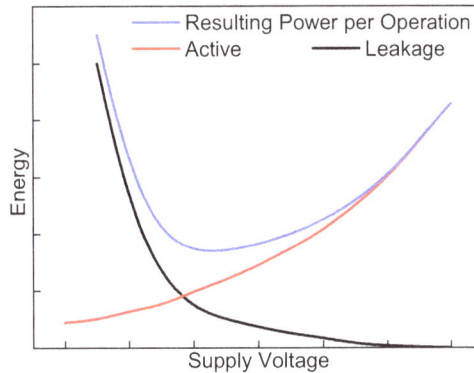

Figure 1. Energy consumption as a function of the supply voltage [10].

With its lower lateral electric field and longer effective channel length, extensionless devices, also known as underlap devices, have demonstrated a better SCE, providing a more scalable structure, better analog performance, and advantages in memory applications [11–15]. Moreover, they have shown a better subthreshold behavior, which in the weak inversion regime is favorable for low power low voltage applications [16].

Originally, the dynamic threshold (DT) concept was explored in partially depleted SOI (PDSOI) devices as a way to avoid the floating body problems. The gate and the body are short-circuited; consequently, the body is never floating [17]. However, the bulk-drain junction can be forward biased if the gate voltage is higher than 0.7 V, which is a disadvantage for these devices. Recently, with the advent of UTBB devices, a new generation of dynamic threshold voltage configuration (DT2) has been studied [3,18]. In this case, the front gate is connected to the back-gate and the same concept is achieved: during the V_G sweep, the threshold voltage (V_T) is dynamically reduced, as it is a function of the V_{Body} (in PDSOI) or V_B (in UTBB), which is equal to V_G, improving the device performance [17].

Some optimizations have been reported about this approach, regarding the ground plane and the channel length influence in these operation modes [19] or the impact of the silicon film thickness [20]. The effect of the buried oxide thickness on the analog figures of merits has been studied [20–23], as well as the circuit for generating the back gate of the eDT mode for a sleep transistor, taking into account the back gate capacitance influence on the circuit performance. Finally, the application of the dynamic threshold technique in UTBB devices as a power switch has also been discussed [24].

Therefore, the goal of this work is to compare the impact of the underlap length (L_{UL}), including the self-aligned devices (L_{UL} = 0 nm), on UTBB SOI MOSFETs when submitted to a conventional and a dynamic threshold voltage (DT2) configuration, focusing on low power low voltage applications. In order to enhance the back gate influence on V_T, a back gate bias with a multiple value of the front gate voltage ($V_B = k \times V_G$) is also used [3,18], with $k > 1$, which is called the eDT mode in this paper.

2. Device Description

Figure 2 shows the structure of an extensionless UTBB FDSOI device. The devices were built on 300 mm diameter SOI wafers with a 20 nm silicon film (t_{Si}) and 10 nm buried oxide (t_{BOX}). A B-implantation was performed for GP doping. The gate stack is composed by 5 nm SiO_2 and a 5 nm plasma enhanced atomic layer deposition (PEALD) TiN layer capped with 100 nm a-Si. A low energy As-implantation was performed on the standard nMOSFETs (self-aligned with lightly-doped drain—LDD) for the extensions while extensions were left free for the extensionless ones. This was followed by a formation of a first nitride spacer of 15 nm or 20 nm width and of the implanted Si-epitaxial raised Source-Drain (SEG) As-HDD (highly-doped drain) [25]. After the HDD a second nitride spacer is formed followed by silicidation and a standard back-end of line process. The effective channel length (L_{eff}) and width (W) were fixed to 70 nm and 920 nm, respectively. The underlap lengths (L_{UL}) are 0 nm (for the self-aligned devices with LDD), or the first spacer width (15 nm or 20 nm).

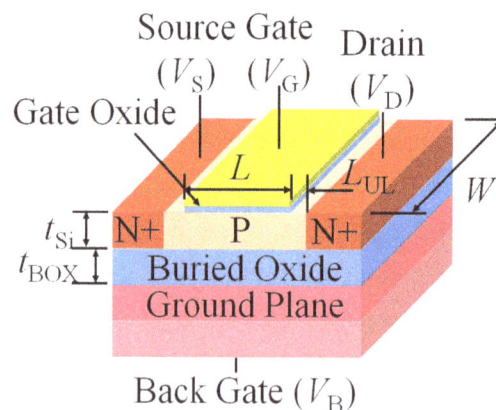

Figure 2. Extensionless UTBB FDSOI (ultrathin-body-and-buried-oxide fully-depleted-silicon-on-insulator) structure considered in this work.

It is important to mention that devices with an underlap of 10 nm have been analyzed in [26]. They presented a worse behavior than the self-aligned devices, due to the shorter effective channel length. These devices have a larger difference between the doping concentration of the channel and the source and drain regions, which increase the lateral diffusion [26]. Therefore, this case will not further be discussed here.

3. Results and Discussion

The I_D-V_G (drain current as a function of front gate voltage) characteristics are presented in Figure 3 for the self-aligned case and for an underlap of 20 nm in the conventional and the eDT ($k = 3$) modes.

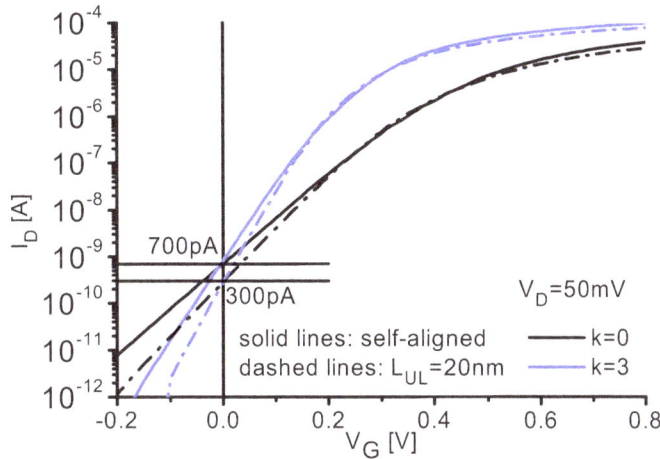

Figure 3. Drain current *versus* front gate voltage for self-aligned and an underlap length of 20nm in conventional and eDT ($k = 3$) operation. $L_{eff} = 70$ nm, $W_{eff} = 920$ nm, $t_{Si} = 14$ nm, $t_{BOX} = 18$ nm.

Table 1. I_{ON}/I_{OFF} ratio for self-aligned and an underlap length of 20 nm and for conventional and eDT ($k = 3$) configurations. I_{OFF} @ $V_G = 0$V and I_{ON} @ $V_{GT} = 200$ mV. $L_{eff} = 70$ nm, $W_{eff} = 920$ nm, $t_{Si} = 14$ nm, $t_{BOX} = 18$ nm.

Self-Aligned, $k = 0$	Self-Aligned, $k = 3$	$L_{UL} = 20$ nm, $k = 0$	$L_{UL} = 20$ nm, $k = 3$
3.08×10^4	6.54×10^4	6.48×10^4	10.52×10^4

One can observe in Figure 3 a lower I_{OFF} (at $V_G = 0$V), *i.e.*, a lower off-state current, for underlap devices in the conventional mode. A lower off-current is aimed in order to reduce the power dissipation during off-state. This behavior can be due to the longer L_{eff}, and better SCE. This remains in the eDT mode, since the back gate bias applied in both cases is the same. As the devices reach weak inversion, the L_{eff} approaches L_G and the I_D of the underlap device becomes closer to the self-aligned counterpart [15]. In strong inversion, the I_D decreases around 10% due to the higher total resistance of the underlap devices, but I_{OFF} reduces around 57% [14]. Also, when the back gate bias is increased (in DT and eDT conditions) the threshold voltage reduces dynamically and a higher I_{ON} can be achieved. This results in a higher I_{ON}/I_{OFF} ratio for underlap devices in eDT mode (Table 1).

The subthreshold swing (S) and the maximum transconductance were extracted and are presented as a function of the k-values for three underlap lengths in Figures 4 and 5.

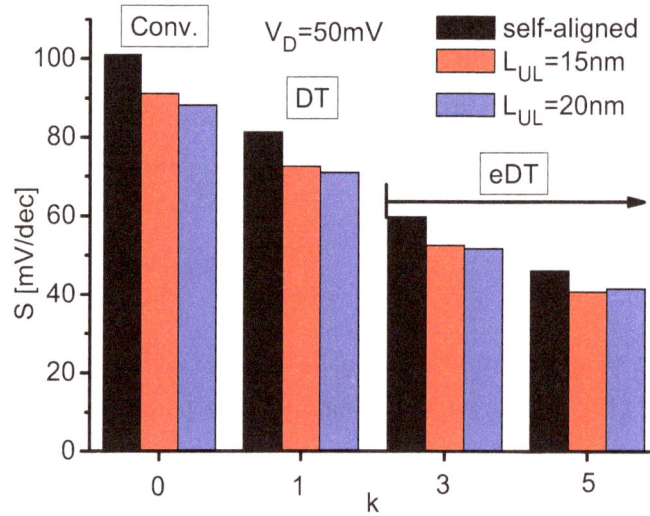

Figure 4. Subthreshold swing for various k-values and underlap lengths (based on reference [27]). L_{eff} = 70 nm, W_{eff} = 920 nm, t_{Si} = 14 nm, t_{BOX} = 18 nm.

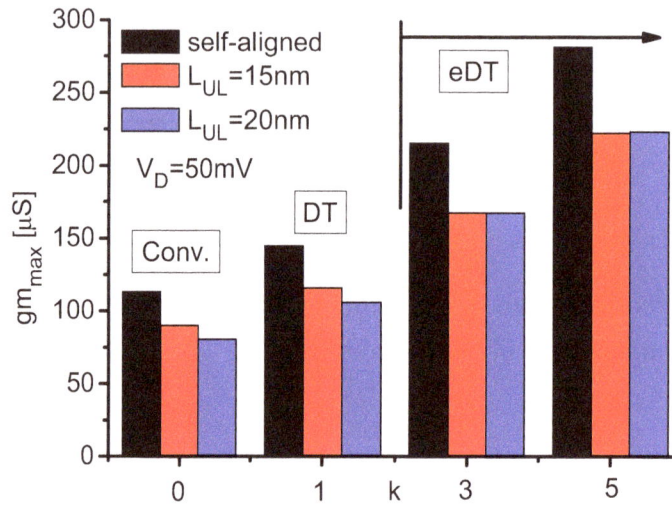

Figure 5. Maximum transconductance for various k-values and underlap lengths (based on reference [27]). L_{eff} = 70 nm, W_{eff} = 920 nm, t_{Si} = 14 nm, t_{BOX} = 18 nm.

The higher I_D variation with V_G shown in Figure 1 for underlap devices, and now also with V_B for higher k values, leads to a lower subthreshold swing, which is important for low power devices by enabling reduced supply voltages [28]. Moreover, although the underlap devices present a lower maximum transconductance due to the higher total resistance, the eDT mode can overcome this drawback.

Another important aim for low power applications is the reduction of the threshold voltage without raising the off-current, since it means a lower supply voltage added to a constant, or lower, leakage current Equation (1) [10]. Table 2 and Figure 6 show, respectively, the absolute and normalized threshold voltage for the three devices studied and for various k-values.

As expected, the threshold voltage decreases for higher k values due to the stronger DT effect (Table 2). This effect can also be strengthened by the lower lateral electric field for a higher underlap, which is seen by a lower normalized threshold voltage for DT and eDT in Figure 6.

Table 2. Threshold voltage for various k-values and underlap lengths (based on reference [27]). $L_{\text{eff}} = 70$ nm, $W_{\text{eff}} = 920$ nm, $t_{\text{Si}} = 14$ nm, $t_{\text{BOX}} = 18$ nm.

k	Threshold Voltage [V]		
	Self-Aligned	$L_{\text{UL}} = 15$ nm	$L_{\text{UL}} = 20$ nm
0	0.45	0.48	0.48
1	0.40	0.40	0.37
3	0.29	0.30	0.25
5	0.22	0.24	0.20

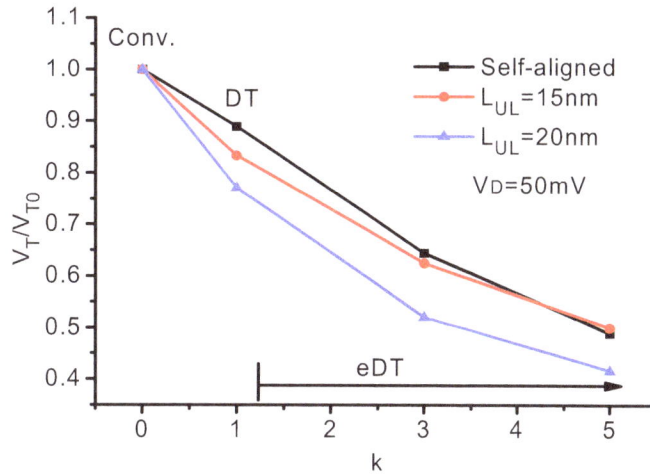

Figure 6. Normalized threshold voltage for various k-values and underlap lengths (based on reference [27]). $L_{\text{eff}} = 70$ nm, $W_{\text{eff}} = 920$ nm, $t_{\text{Si}} = 14$ nm, $t_{\text{BOX}} = 18$ nm.

Concerning the leakage current, there is also the gate-induced-drain-leakage (GIDL), which can be seen in Figure 7 for various k values as a I_{D}-V_{GT} (drain current *versus* gate overdrive, where $V_{\text{GT}} = V_{\text{G}}-V_{\text{T}}$) characteristics (left) and for the three devices studied (right).

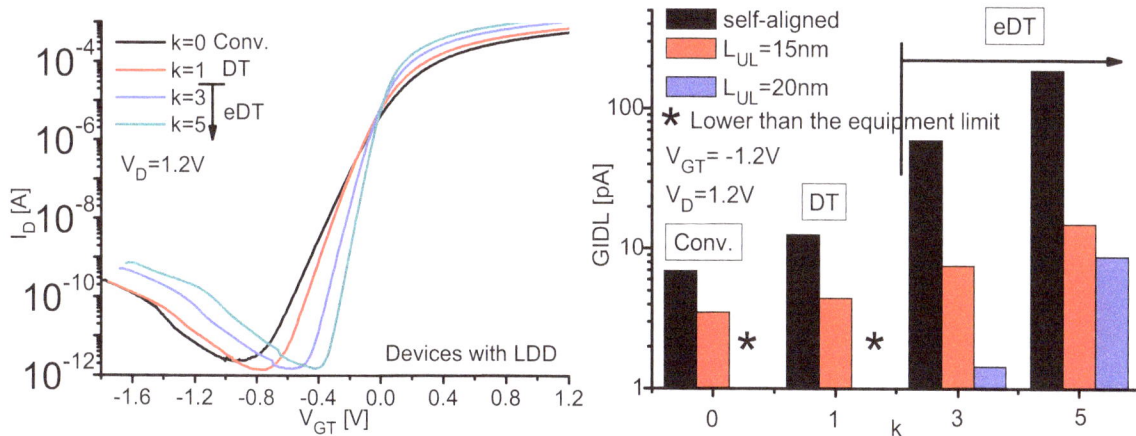

Figure 7. Drain current *versus* overdrive voltage for $V_{\text{D}} = 1.2$ V for various k-values (**left**) and gate-induced-drain-leakage for different k-factors and underlap lengths (**right**) (based on reference [27]). $L_{\text{eff}} = 70$ nm, $W_{\text{eff}} = 920$ nm, $t_{\text{Si}} = 14$ nm, $t_{\text{BOX}} = 18$ nm.

Concerning the leakage current, as it has already been shown before, I_{OFF} (at $V_{\text{G}} = 0$V) is lower in underlap devices and is the same in DT and eDT modes. Besides, it is possible to see, in Figure 7, a higher GIDL in DT and eDT modes, most probably due to a lower potential induced by the lower V_B. For higher k-values, the difference between the front and back gate potential increases, enhancing the transversal electric field. This can generate more tunneling charges near the drain, worsening this leakage. However, underlap devices reduce drastically this leakage by a lower lateral electric field [29]. One can notice from the right side of Figure 7, for example, that the 20 nm-underlap for $k = 5$ presents almost the same value than the self-aligned counterpart in the conventional mode.

Figure 8 shows the transistor efficiency, *i.e.*, the transconductance over drain current ratio (gm/I_D) for various k values and for the three underlap lengths in weak inversion, at a normalized $I_D/(W/L)$ of 1 nA.

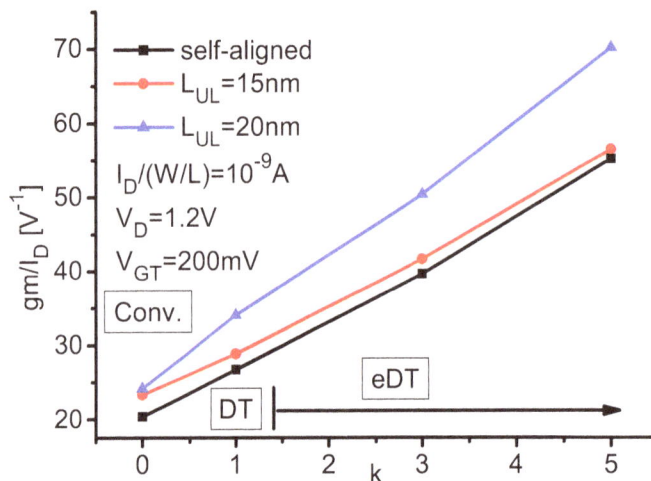

Figure 8. Transconductance over drain current ratio in weak inversion for various k-values and underlap lengths at a normalized drain current of 1nA (based on reference [27]). L_{eff} = 70 nm, W_{eff} = 920 nm, t_{Si} = 14 nm, t_{BOX} = 18 nm.

A higher efficiency is observed for longer underlap and higher k values, due to the lower subthreshold swing (Figure 4). In other words, one can achieve a higher amplification with the same supply energy in underlap devices in DT and eDT modes.

The intrinsic voltage gain of the studied devices is given for various k values in Figure 9.

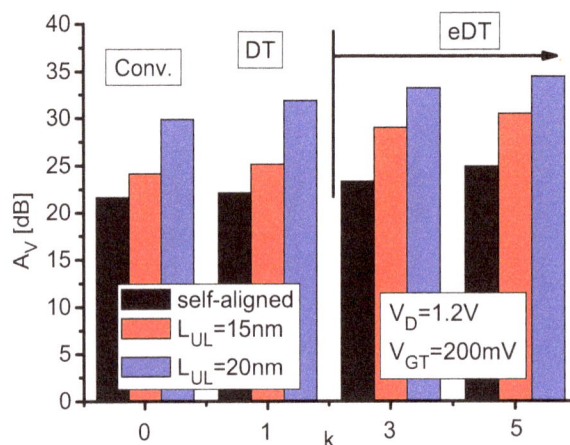

Figure 9. Intrinsic voltage gain for various k-values and underlap lengths (based on reference [27]). L_{eff} = 70 nm, W_{eff} = 920 nm, t_{Si} = 14 nm, t_{BOX} = 18 nm.

One observes an increased intrinsic voltage gain ($|A_V|$) for longer underlap and higher k-values. Since $|A_V| = V_{EA} \times gm/I_D$, Figure 10 shows the Early voltage (V_{EA}) and the gm/I_D in strong inversion for various k values and underlap lengths.

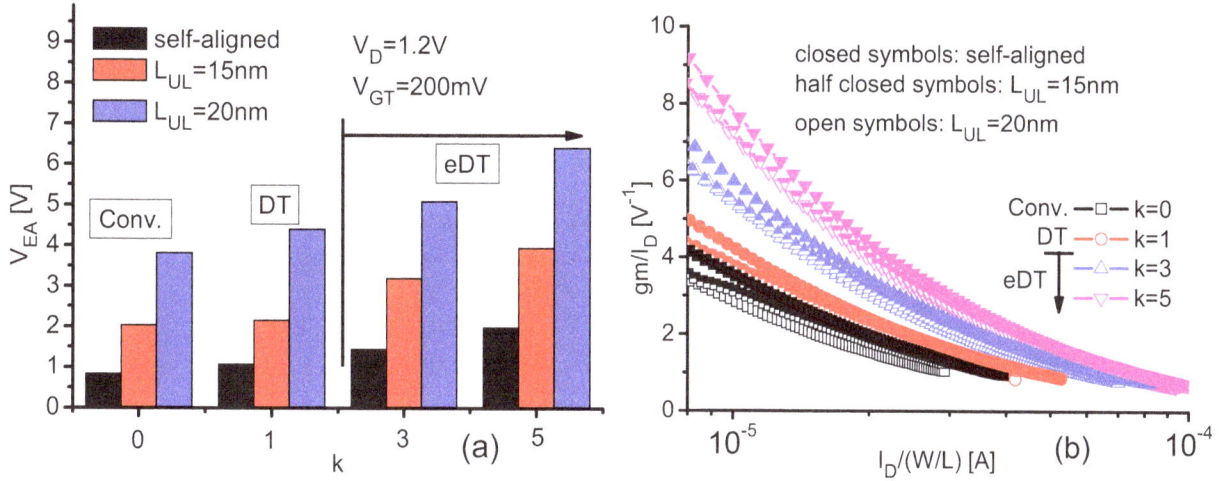

Figure 10. Early voltage (**a**) and transconductance over drain current ratio in strong inversion (**b**) for various k-values and underlap lengths (based on reference [27]). $L_{eff} = 70$ nm, $W_{eff} = 920$ nm, $t_{Si} = 14$ nm, $t_{BOX} = 18$ nm.

Although there is a slight reduction of the gm/I_D in strong inversion for longer underlap (Figure 10b) due to the higher total resistance [12,27], it is negligible when compared to the improvement of V_{EA} (Figure 10a), thanks to the lower lateral electric field [12,29]. Therefore, the main reason for the $|A_V|$ tendency is the substantial reduction of the lateral electric field and, consequently, a higher Early voltage (Figure 10a). Also, a higher influence of the transversal electric field can improve this parameter in DT and eDT operations.

Regarding the k values, both, the gm/I_D and V_{EA}, increase for higher k values in strong inversion (Figure 10a,b).

About the gm/I_D in strong inversion, as it is a drawback for longer underlap, one can observe that, for example, the self-aligned transistor in the conventional mode presents almost the same value than the 20 nm-underlap device for $k = 1$. This means that the eDT mode is overcoming the undesired trend.

In order to better evaluate the correlation between the lateral and the transversal electric field, Figure 11 shows the DIBL (Drain Induced Barrier Lowering) for devices with 20 nm and 15 nm underlap in the 3 operation modes considered.

In this figure, the DIBL can represent an estimation of the drain electric field existing in the channel. On the other hand, the k-values can be seen as operating on the transversal electric field, since a higher k-value leads to a higher difference between the front and the back interface potential.

One can observe a higher variation for the 20 nm-underlap case than for 15 nm when a higher k value is applied. It means that the longer underlap, which corresponds with a lower lateral electric field, is more susceptible to the dynamic threshold effect. In other words, the higher transversal electric field, resulting from a higher k-factor, strongly acts when the lateral electric field is lower.

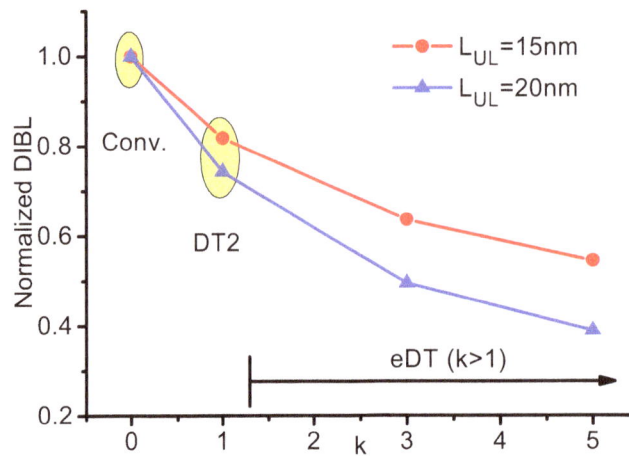

Figure 11. Normalized DIBL as a function of k-value for an underlap of 15 nm and 20 nm (based on reference [27]). L_{eff} = 70 nm, W_{eff} = 920 nm, t_{Si} = 14 nm, t_{BOX} = 18 nm.

4. Conclusions

The impact of the underlap length, including the self-aligned LDD device, on the main parameters was analyzed for three operation conditions: the conventional (V_B = 0 V), the standard dynamic threshold (DT, V_B = V_G) and the enhanced DT (eDT, V_B = kV_G), focusing on low power applications.

For low power low voltage applications, the best results were $S \cong$ 41 mV/dec, V_T = 0.2 V, $gm/I_D \approx$ 70 V^{-1} (in weak inversion) and $|A_V| \approx$ 34 dB for the 20 nm-underlap device, a channel length of 70 nm and in eDT mode.

Although the longer underlap presented a lower ON-current and transconductance due to the higher total resistance, the dynamic threshold reduction compensates these drawbacks. On the other hand, a longer underlap substantially diminishes the GIDL current, which is a negative point of the dynamic threshold and enhanced modes, since it also means a more negative back gate bias. Looking at the OFF-current, the I_{ON}/I_{OFF} ratio, the subthreshold swing, the threshold voltage, the gm/I_D ratio and the intrinsic voltage gain, both features, longer underlap and eDT, mode improve these parameters.

Moreover, longer underlap devices are more susceptible with the increase of the k-factor, leading to a further performance improvement. This can be explained by the lower lateral electric field and higher influence of the transversal one, which strengthen the dynamic threshold effect.

Thus, the eDT mode or underlap devices alone present important shortcomings for low power low voltage applications, such as higher GIDL for eDT mode and lower I_{ON} and gm/I_D ratio for underlap devices. However, combined, they compensate the drawback of the other feature. The 20 nm-underlap device in eDT mode with k = 5 presented about the same GIDL current than the self-aligned transistor in the conventional mode. And even for k = 3, the eDT mode improves the lower I_{ON} and gm/I_D ratio shown by the longer underlap.

Therefore, the 20 nm-underlap device in eDT mode with k = 5 presented the best performance for low power low voltage applications, by enabling a low supply voltage together with a low leakage current.

Acknowledgments

The authors acknowledge FAPESP (Fundação de Amparo à Pesquisa do Estado de Sao Paulo), CAPES (Coordenação de Aperfeiçoamento de Pessoal de Nível Superior), CNPq (Conselho Nacional de Desenvolvimento Científico e Tecnológico) and FWO (Fonds Wetenschappelijke Onderzoek (FWO)—Vlaanderen) for the financial support.

Author Contributions

Katia Sasaki measured and discussed the results. Marc Aoulaiche, Eddy Simoen and Cor Claeys fabricated the devices. Joao Martino discussed and coordinated the overall research.

Conflicts of Interest

The authors declare no conflict of interest.

References

1. Fenouillet-Beranger, C.; Denorme, S.; Perreau, P.; Buj, C.; Faynot, O.; Andrieu, F.; Tosti, L.; Barnola, S.; Salvetat, T.; Garros, X.; *et al.* FDSOI devices with thin BOX and ground plane integration for 32 nm node and below. *Solid State Electron.* **2009**, *53*, 730–734.
2. Fujiwara, M.; Morooka, T.; Yasutake, N.; Ohuchi, K.; Aoki, N.; Tanimoto, H.; Kondo, M.; Miyano, K.; Inaba, S.; Ishimaru, K.; *et al.* Impact of BOX scaling on 30 nm gate length FD SOI MOSFET. In Proceedings of the 2005 IEEE International SOI Conference, Honolulu, HI, USA, 3–6 October 2005; pp. 180–182.
3. Kilchytska, V.; Flandre, D.; Andrieu, F. On The UTBB SOI MOSFET performance improvement in Quasi-Double-Gate Regime. In Proceedings of the ESSDERC—European Solid-State Device Conference, Bordeaux, France, 17–21 September 2012; p. 246.
4. Nguyen, B.-Y.; Maleville, C. Advanced Substrate News, 2014. Available online: http//www. advancedsubstratenews.com/2014/03/fd-soi-back-to-basics-for-best-cost-energy-efficiency-and-performance (accessed on 7 May 2014).
5. Yan, R.; Duane, R.; Razavi, P.; Afzalian, A.; Ferain, I.; Lee, C.W.; Dehdashti-Akhavan, N.; Bourdelle, K.; Nguyen, B.Y.; Colinge, J.P. LDD Depletion Effects in thin-BOX FDSOI Devices with a Ground Plane. In Proceedings of the IEEE International SOI Conference, Foster City, CA, USA, 5–8 October 2009.
6. Nakamura, S.; Kawasaki, J.; Kumagai, Y.; Usami, K. Measurements of the Minimum Energy Point in Silicon-on-Thin-BOX (SOTB) and Bulk MOSFET. In Proceedings of the Joint International EUROSOI Workshop and International Conference on Ultimate Integration on Silicon, Bologna, Italy, 26–28 January 2015; pp. 193–196.
7. Ohtou, T.; Saraya, T.; Hiramoto, T. Variable-body-factor SOI MOSFET with ultrathin buried oxide for adaptive threshold voltage and leakage control. *IEEE Trans. Electron. Devices* **2008**, *55*, 40–47.

8. Yanagi, S.; Nakakubo, A.; Omura, Y. Proposal of partial-ground-plane (PGP) silicon-on-insulator (SOI) MOSFET for deep sub-0.1 μm channel regime. *IEEE Electron. Device Lett.* **2001**, *22*, 278–280.

9. Xiong, W.; Colinge, J.P. Self-aligned ground-plane fully depleted SOI MOSFET. *Electron. Lett.* **1999**, *35*, 2059–2060.

10. Kimura, S. Overview of ultra-low power devices, focus on SOTB. In Proceedings of the IEEE SOI-3D-Subthreshold Microelectronics Technology Unified Conference (S3S), San Francisco, CA, USA, 6–9 October 2014.

11. Trivedi, V.; Fossum, J.G.; Chowdhury, M.M. Nanoscale FinFETs with Gate-Source/Drain Underlap. *IEEE Trans. Electron. Devices* **2005**, *52*, 56–62.

12. Santos, S.; Nicoletti, T.; Martino, J.A. Analog Performance of Gate-Source/Drain. Underlap Triple Gate SOI nMOSFET. *ECS Trans.* **2011**, *39*, 239–246.

13. Song, K.-W.; Jeong, H.; Lee, J.-W.; Hong, S.I.; Tak, N.-K.; Kim, Y.-T.; Choi, Y.L.; Joo, H.S.; Kim, S.H.; Song, H.J.; *et al.* 55 nm Capacitorless. 1T-DRAM Cell. Transistor with non-overlap structure. In Proceedings of the IEEE International Electron Devices Meeting, San Francisco, CA, USA, 15–17 December 2008; pp. 1–4.

14. Nicoletti, T.; Santos, S.; Almeida, L.; Martino, J.; Aoulaiche, M.; Veloso, A.; Jurczak, M.; Simoen, E.; Claeys, C. The Impact of Gate Length Scaling on UTBOX FDSOI Devices: The Digital/Analog Performance of Extension-less Structures. In Proceedings of the 13th Ultimate Integration on Silicon, Grenoble, France, 6–7 March 2012; p. 161.

15. Fossum, J.G.; Chowdhury, M.M.; Trivedi, V.P.; King, T.-J.; Choi, Y.-K.; An, J.; Yu, B. Physical Insights on Design and Modeling of Nanoscale FinFETs. In Proceedings of the IEEE International Electron Devices Meeting, Washington, DC, USA, 8–10 December 2003.

16. Vitale, S.A.; Peter, W.W.; Checka, N.; Kedzierski, J.; Keast, C.L. FDSOI Process Technology Subthreshold Operation Ultra-Low-Power Electronics. In Proceedings of the 219th ECS Meeting, Montreal, QC, Canada, 1–6 May 2011; pp. 179–188.

17. Colinge, J.P. An SOI voltage-controlled bipolar-MOS device. *IEEE Trans. Electron. Devices* **1987**, *34*, 845–849.

18. Sasaki, K.R.A.; Manini, M.B.; Martino, J.A.; Aoulaiche, M.; Simoen, E.; Witters, L.; Claeys, C. Ground plane influence on enhanced dynamic threshold UTBB SOI nMOSFETs. In Proceedings of 9th ICCDCS Conference, Playa del Carmen, Mexico, 2–4 April 2014; pp. 1–4.

19. Sasaki, K.R.A.; Manini, M.B.; Simoen, E.; Claeys, C.; Martino, J.A. Enhanced dynamic threshold voltage UTBB SOI nMOSFETs. *Solid-State Electron.* **2015**, in press.

20. Sasaki, K.R.A.; Aoulaiche, M.; Simoen, E.; Claeys, C.; Martino, J.A. Silicon Film Thickness Influence on Enhanced Dynamic Threshold UTBB SOI nMOSFETs. In Proceedings of the SBMicro 2014, 29th Symposium on Microelectronics Technology and Devices, Aracaju, SE, Brazil, 1–5 September 2014; pp. 1–4.

21. Kilchytska, V.; Bol, D.; de Vos, J.; Andrieu, F.; Flandre, D. Quasi-double gate regime to boost UTBB SOI MOSFET performance in analog and sleep transistors applications. *Solid State Electron.* **2013**, *84*, 28–37.

22. Bol, D.; Kilchytska, V.; de Vos, J.; Andrieu, F.; Flandre, D. Quasi-Double Gate Mode for Sleep transistors in UTBB FD SOI Low-Power High Speed Applications. SOI Conference (SOI). In Proceedings of the 2012 International, Napa, CA, USA, 1–4 October 2012; pp. 1–2.

23. Arshad, M.K.M.; Kilchytska, V.; Makovejev, S.; Olsen, S.; Andrieu, F.; Raskin, J.P.; Flandre, D. UTBB SOI MOSFETs Analog Figures of Merits: Effect of Ground Plane and Asymmetric Double Gate Regime. In Proceedings of the EUROSOI, Montpellier, France, 23–25 January 2012; pp. 111–112.

24. Le Coz, J.; Pelloux-Prayer, B.; Giraud, B.; Giner, F.; Flatresse, P. DTMOS Power Switch in 28 nm UTBB FD-SOI Technology. In Proceedings of the IEEE SOI-3D-Subthreshold Microelectronics Technology Unified Conference (S3S), Monterey, CA, USA, 7–10 October 2013; pp. 1–2.

25. Nicoletti, T.; Santos, S.D.; Sasaki, K.R.A.; Martino, J.A.; Aoulaiche, M.; Simoen, S.; Claeys, C. The Activation Energy Dependence on the Electric Field in UTBOX SOI FBRAM Devices. In Proceedings of the IEEE SOI-3D-Subthreshold Microelectronics Technology Unified Conference (S3S), Monterey, CA, USA, 7–10 October 2013; pp. 1–2.

26. Santos, S.D.; Nicoletti, T.; Aoulaiche, M.; Martino, J.A.; Veloso, A.; Jurczak, M.; Simoen, E.; Claeys, C. In ECS Transactions, Spacer Length and Tilt Implantation Influence on Scaled UTBOX FD MOSFETs. In Proceedings of the SBMicro 2012: 27th Symposium on Microelectronics Technology and Devices, 30 August–2 September 2012; pp. 483–489.

27. Sasaki, K.R.A.; Aoulaiche, M.; Simoen, E.; Claeys, C.; Martino, J.A. Influence of Underlap on UTBB SOI MOSFETs in Dynamic Threshold Mode. In Proceedings of the IEEE SOI-3D-Subthreshold Microelectronics Technology Unified Conference (S3S), San Francisco, CA, USA, 6–9 October 2014; pp. 1–3.

28. Woo, J. Tunnel-FETs. In Proceedings of IEEE SOI-3D-Subthreshold Microelectronics Technology Unified Conference (S3S), San Francisco, CA, USA, 6–9 October 2014.

29. Sasaki, K.R.A.; Nicoletti, T.; Almeida, L.; Santos, S.D.; Nissimoff, A.; Aoulaiche, M.; Simoen, E.; Claeys, C.; Martino, J.A. Improved retention times in UTBOX nMOSFETs for 1T-DRAM applications. *Solid State Electron.* **2014**, *97*, 30–37.

Permissions

List of Contributors

Arthur Spivak
The VLSI Systems Center, LPCAS, Ben-Gurion University, P.O.B. 653, Be'er-Sheva 84105, Israel

Alexander Belenky
The VLSI Systems Center, LPCAS, Ben-Gurion University, P.O.B. 653, Be'er-Sheva 84105, Israel

Alexander Fish
Department of Electrical and Computer Engineering, Bar Ilan University, Ramat Gan, 52100, Israel

Orly Yadid-Pecht
Department of Electrical Engineering, University of Calgary, Alberta T2N 1N4, Canada

Suma George
Georgia Institute of Technology, Atlanta 30363, GA, USA

Jennifer Hasler
Georgia Institute of Technology, Atlanta 30363, GA, USA

Scott Koziol
Georgia Institute of Technology, Atlanta 30363, GA, USA

Stephen Nease
Georgia Institute of Technology, Atlanta 30363, GA, USA

Shubha Ramakrishnan
Georgia Institute of Technology, Atlanta 30363, GA, USA

Shin-Chi Lai
Department of Electrical Engineering, National Cheng Kung University, 701 Tainan, Taiwan

Yueh-Shu Lee
Department of Electrical Engineering, National Cheng Kung University, 701 Tainan, Taiwan

Sheau-Fang Lei
Department of Electrical Engineering, National Cheng Kung University, 701 Tainan, Taiwan

Bo Liu
Electronic System Group, Department of Electrical Engineering, Technische Universiteit Eindhoven, Den Dolech 2, 5612AZ, Eindhoven, The Netherlands
Holst Centre/Imec-nl, High Tech Campus 31, 5656AE, Eindhoven, The Netherlands

Jose Pineda de Gyvez
Electronic System Group, Department of Electrical Engineering, Technische Universiteit Eindhoven, Den Dolech 2, 5612AZ, Eindhoven, The Netherlands

Maryam Ashouei
Holst Centre/Imec-nl, High Tech Campus 31, 5656AE, Eindhoven, The Netherlands

Qing Gao
Department of Electrical and Computer Engineering, University of Calgary, AB T2N1N4, Canada

Orly Yadid-Pecht
Department of Electrical and Computer Engineering, University of Calgary, AB T2N1N4, Canada

Chika Ofili
Integrated Sensors, Intelligent Systems (ISIS) Laboratory, Electrical and Computer Engineering Department, University of Calgary, Calgary, AB T2N 1N4, Canada

Stanislav Glozman
VLSI Systems Center, Ben-Gurion University of the Negev, POB 653, Beer-Sheva 84105, Israel

Orly Yadid-Pecht
Integrated Sensors, Intelligent Systems (ISIS) Laboratory, Electrical and Computer Engineering Department, University of Calgary, Calgary, AB T2N 1N4, Canada

Pooja Batra
IBM Corporation Systems and Technology Group, Hopewell Junction, NY 12533, USA

Spyridon Skordas
IBM Corporation Systems and Technology Group, Albany, NY 12203, USA

Douglas LaTulipe
Formerly with IBM Corporation Systems and Technology Group

Kevin Winstel
IBM Corporation Systems and Technology Group, Albany, NY 12203, USA

Chandrasekharan Kothandaraman
IBM Corporation Systems and Technology Group, Hopewell Junction, NY 12533, USA

Ben Himmel
Formerly with IBM Corporation Systems and Technology Group

Gary Maier
IBM Corporation Systems and Technology Group, Hopewell Junction, NY 12533, USA

Bishan He
IBM Corporation Systems and Technology Group, Hopewell Junction, NY 12533, USA

Deepal Wehella Gamage
IBM Corporation Systems and Technology Group, Hopewell Junction, NY 12533, USA

John Golz
IBM Corporation Systems and Technology Group, Hopewell Junction, NY 12533, USA

Wei Lin
IBM Corporation Systems and Technology Group, Albany, NY 12203, USA

Tuan Vo
Formerly with IBM Corporation Systems and Technology Group

Deepika Priyadarshini
IBM Corporation Systems and Technology Group, Albany, NY 12203, USA

Alex Hubbard
IBM Corporation Systems and Technology Group, Albany, NY 12203, USA

Kristian Cauffman
IBM Corporation Systems and Technology Group, Albany, NY 12203, USA

Brown Peethala
IBM Corporation Systems and Technology Group, Albany, NY 12203, USA

John Barth
Formerly with IBM Corporation Systems and Technology Group

Toshiaki Kirihata
IBM Corporation Systems and Technology Group, Hopewell Junction, NY 12533, USA

Troy Graves-Abe
IBM Corporation Systems and Technology Group, Hopewell Junction, NY 12533, USA

Norman Robson
IBM Corporation Systems and Technology Group, Hopewell Junction, NY 12533, USA

Subramanian Iyer
IBM Corporation Systems and Technology Group, Hopewell Junction, NY 12533, USA

Bennion Redd
Department of Electrical & Computer Engineering, University of Utah, 1692 Warnock Engineering Bldg., 72 S. Central Campus Dr., Salt Lake City, UT 84112

Spencer Kellis
Division of Biology and Biological Engineering, California Institute of Technology, Mail Code 216-76, 1200 E, California Blvd., Pasadena, CA 91125, USA

Nathaniel Gaskin
701 E Charleston Rd, Palo Alto, CA 94303, USA

Richard Brown
Department of Electrical & Computer Engineering, University of Utah, 1692 Warnock Engineering Bldg., 72 S. Central Campus Dr., Salt Lake City, UT 84112

Jonathan F. Bolus
Department of Electrical & Computer Engineering, University of Virginia, 351 McCormick Rd., Charlottesville, VA 22904, USA

Benton H. Calhoun
Department of Electrical & Computer Engineering, University of Virginia, 351 McCormick Rd., Charlottesville, VA 22904, USA

Travis N. Blalock
Department of Electrical & Computer Engineering, University of Virginia, 351 McCormick Rd., Charlottesville, VA 22904, USA

Ro'ee Eitan
Intel Israel LTD., S.B.I. Park Har Hotzvim, CFF8, Jerusalem 91031, Israel

Ariel Cohen
Intel Israel LTD., S.B.I. Park Har Hotzvim, CFF8, Jerusalem 91031, Israel

Katia Regina Akemi Sasaki
LSI/PSI — Integrated Systems Laboratory/Department of Electronic Systems Engineering,
University of Sao Paulo, Av. Prof. Luciano Gualberto, trav.3, n.158, Sao Paulo 05508-010, Brazil

Marc Aoulaiche
IMEC — Interuniversity Microelectronic Centre, Kapeldreef 75, B-3001 Leuven, Belgium

Eddy Simoen
IMEC — Interuniversity Microelectronic Centre, Kapeldreef 75, B-3001 Leuven, Belgium

Cor Claeys
IMEC — Interuniversity Microelectronic Centre, Kapeldreef 75, B-3001 Leuven, Belgium
Electrical Engineering Department, KU Leuven, Kasteelpark Arenberg 10, B-3001 Leuven, Belgium

Joao Antonio Martino
LSI/PSI — Integrated Systems Laboratory/Department of Electronic Systems Engineering,
University of Sao Paulo, Av. Prof. Luciano Gualberto, trav.3, n.158, Sao Paulo 05508-010, Brazil